Quantities, Units and Symbols in Physical Chemistry

INTERNATIONAL UNION OF PURE AND APPLIED CHEMISTRY
PHYSICAL CHEMISTRY DIVISION
COMMISSION ON PHYSICOCHEMICAL SYMBOLS, TERMINOLOGY AND UNITS

IUPAC

INTERNATIONAL UNION OF PURE AND APPLIED CHEMISTRY

Quantities, Units and Symbols in Physical Chemistry

Prepared
for publication by

IAN MILLS
Reading

TOMISLAV CVITAŠ
Zagreb

KLAUS HOMANN
Darmstadt

NIKOLA KALLAY
Zagreb

KOZO KUCHITSU
Tokyo

SECOND EDITION

OXFORD
BLACKWELL SCIENTIFIC PUBLICATIONS
LONDON EDINBURGH BOSTON
MELBOURNE PARIS BERLIN VIENNA

Blackwell Scientific Publications
Editorial Offices:
Osney Mead, Oxford OX2 0EL
25 John Street, London WC1N 2BL
23 Ainslie Place, Edinburgh EH3 6AJ
238 Main Street, Cambridge
Massachusetts 02142, USA
54 University Street, Carlton
Victoria 3053, Australia

Other Editorial Offices:
Librairie Arnette SA
2, rue Casimir-Delavigne
75006 Paris
France

Blackwell Wissenschafts-Verlag GmbH
Meinekestrasse 4
D-1000 Berlin 15
Germany

Blackwell MZV
Feldgasse 13
A-1238 Wien
Austria

First published 1988
Reprinted 1988
Reprinted as paperback 1989
Russian translation 1988
Hungarian translation 1990
Indian reprint edition 1990
Japanese translation 1991
Second edition 1993

Set by Macmillan India Ltd
Printed and bound in Great Britain
at The University Press, Cambridge

DISTRIBUTORS

Marston Book Services Ltd
PO Box 87
Oxford OX2 0DT
(*Orders*: Tel: 0865 791155
Fax: 0865 791927
Telex: 837515)

Australia
Blackwell Scientific Publications Pty Ltd
54 University Street
Carlton, Victoria 3053
(*Orders*: Tel: (03) 347-5552)

USA and Canada
CRC Press, Inc.
2000 Corporate Blvd, NW
Boca Raton
Florida 33431

A catalogue record for this book
is available from the British Library

ISBN 0-632-03583-8

Library of Congress
Cataloging in Publication Data

Quantities, units and symbols in physical chemistry/
prepared for publication by
Ian Mills . . . [et al.].—2nd ed.
p. cm.
At head of title: International Union
of Pure and Applied Chemistry.
'International Union of Pure
and Applied Chemistry, Physical
Chemistry Division,
Commission on Physicochemical Symbols,
Terminology, and Units'—P. facing t.p.
Includes bibliographical references
and index.
ISBN 0-632-03583-8
1. Chemistry, Physical and
theoretical—Notation.
2. Chemistry,
Physical and theoretical—Terminology.
I. Mills, Ian (Ian M.)
II. International Union of Pure and Applied
Chemistry.
III. International Union of Pure and Applied
Chemistry.
Commission on Physicochemical Symbols,
Terminology, and Units.
QD451.5.Q36 1993
541.3'014—dc20

Contents

Preface

The objective of this manual is to improve the international exchange of scientific information. The recommendations made to achieve this end come under three general headings. The first is the use of quantity calculus for handling physical quantities, and the general rules for the symbolism of quantities and units, described in chapter 1. The second is the use of internationally agreed symbols for the most frequently used quantities, described in chapter 2. The third is the use of SI units wherever possible for the expression of the values of physical quantities; the SI units are described in chapter 3.

Later chapters are concerned with recommended mathematical notation (chapter 4), the present best estimates of physical constants (chapters 5 and 6), conversion factors between SI and non-SI units with examples of their use (chapter 7) and abbreviations and acronyms (chapter 8). References (on p. 133) are indicated in the text by numbers (and letters) in square brackets.

We would welcome comments, criticism, and suggestions for further additions to this book. Offers to assist in the translation and dissemination in other languages should be made in the first instance either to IUPAC or to the Chairman of the Commission.

We wish to thank the following colleagues, who have contributed significantly to this edition through correspondence and discussion:

R.A. Alberty (Cambridge, Mass.); M. Brezinšćak (Zagreb); P.R. Bunker (Ottawa); G.W. Castellan (College Park, Md.); E.R. Cohen (Thousand Oaks, Calif.); A. Covington (Newcastle upon Tyne); H.B.F. Dixon (Cambridge); D.H. Everett (Bristol); M.B. Ewing (London); R.D. Freeman (Stillwater, Okla.); D. Garvin (Washington, DC); G. Gritzner (Linz); K.J. Laidler (Ottawa); J. Lee (Manchester); I. Levine (New York, NY); D.R. Lide (Washington, DC); J.W. Lorimer (London, Ont.); R.L. Martin (Melbourne); M.L. McGlashan (London); J. Michl (Austin, Tex.); K. Niki (Yokohama); M. Palmer (Edinburgh); R. Parsons (Southampton); A.D. Pethybridge (Reading); P. Pyykkö (Helsinki); M. Quack (Zürich); J.C. Rigg (Wageningen); F. Rouquérol (Marseille); G. Schneider (Bochum); N. Sheppard (Norwich); K.S.W. Sing (London); G. Somsen (Amsterdam); H. Suga (Osaka); A. Thor (Stockholm); D.H. Whiffen (Stogursey).

Commission on Physicochemical Symbols,
Terminology and Units

Ian Mills
Tomislav Cvitaš
Klaus Homann
Nikola Kallay
Kozo Kuchitsu

Historical introduction

The *Manual of Symbols and Terminology for Physicochemical Quantities and Units* [1.a], to which this is a direct successor, was first prepared for publication on behalf of the Physical Chemistry Division of IUPAC by M.L. McGlashan in 1969, when he was chairman of the Commission on Physicochemical Symbols, Terminology and Units (I.1). He made a substantial contribution towards the objective which he described in the preface to that first edition as being 'to secure clarity and precision, and wider agreement in the use of symbols, by chemists in different countries, among physicists, chemists and engineers, and by editors of scientific journals'. The second edition of the manual prepared for publication by M.A. Paul in 1973 [1.b], and the third edition prepared by D.H. Whiffen in 1979 [1.c], were revisions to take account of various developments in the Système International d'Unités (SI), and other developments in terminology.

The first edition of *Quantities, Units and Symbols in Physical Chemistry* published in 1988 [2.a] was a substantially revised and extended version of the earlier editions, with a slightly simplified title. The decision to embark on this project was taken at the IUPAC General Assembly at Leuven in 1981, when D.R. Lide was chairman of the Commission. The working party was established at the 1983 meeting in Lingby, when K. Kuchitsu was chairman, and the project has received strong support throughout from all present and past members of Commission I.1 and other Physical Chemistry Commissions, particularly D.R. Lide, D.H. Whiffen and N. Sheppard.

The extensions included some of the material previously published in appendices [1.d–k]; all the newer resolutions and recommendations on units by the Conférence Générale des Poids et Mesures (CGPM); and the recommendations of the International Union of Pure and Applied Physics (IUPAP) of 1978 and of Technical Committee 12 of the International Organization for Standardization (ISO/TC 12). The tables of physical quantities (chapter 2) were extended to include defining equations and SI units for each quantity. The style of the manual was also slightly changed from being a book of rules towards being a manual of advice and assistance for the day-to-day use of practising scientists. Examples of this are the inclusion of extensive footnotes and explanatory text inserts in chapter 2, and the introduction to quantity calculus and the tables of conversion factors between SI and non-SI units and equations in chapter 7.

The manual has found wide acceptance in the chemical community, it has been translated into Russian [2.b], Hungarian [2.c], Japanese [2.d] and large parts of it have been reproduced in the 71st edition of the *Handbook of Chemistry and Physics* published by CRC Press in 1990.

The present volume is a slightly revised and somewhat extended version of the previous edition. The new revisions are based on the recent resolutions of the CGPM [3]; the new recommendations by IUPAP [4]; the new international standards ISO-31 [5, 6]; some recommendations published by other IUPAC commissions; and numerous comments we have received from chemists throughout the world.

Major changes involved the sections: 2.4 Quantum mechanics and Quantum chemistry, 2.7 Electromagnetic radiation and 2.12 Chemical kinetics, in order to include physical quantities used in the rapidly developing fields of quantum chemical computations, laser physics and molecular beam scattering. A new section 3.9 on Dimensionless quantities has been added in the present edition, as well as a Subject index and a list of Abbreviations and acronyms used in physical chemistry.

The revisions have mainly been carried out by Ian Mills and myself with substantial input from Robert Alberty, Kozo Kuchitsu and Martin Quack as well as from other members of the IUPAC Commission on Physicochemical Symbols, Terminology and Units.

Fraunhofer Institute for
Atmospheric Environmental Research
Garmisch-Partenkirchen
June 1992

Tomislav Cvitaš
Chairman
Commission on Physicochemical
Symbols, Terminology and Units

The membership of the Commission during the period 1963 to 1991, during which the successive editions of this manual were prepared, was as follows:

Titular members

Chairman: 1963–1967 G. Waddington (USA); 1967–1971 M.L. McGlashan (UK); 1971–1973 M.A. Paul (USA); 1973–1977 D.H. Whiffen (UK); 1977–1981 D.R. Lide Jr (USA); 1981–1985 K. Kuchitsu (Japan); 1985–1989 I.M. Mills (UK); 1989– T. Cvitaš (Croatia).

Secretary: 1963–1967 H. Brusset (France); 1967–1971 M.A. Paul (USA); 1971–1975 M. Fayard (France); 1975–1979 K.G. Weil (Germany); 1979–1983 I. Ansara (France); 1983–1985 N. Kallay (Croatia); 1985–1987 K.H. Homann (Germany); 1987–1989 T. Cvitaš (Croatia); 1989–1991 I.M. Mills (UK); 1991– M. Quack (Switzerland).

Members: 1975–1983 I. Ansara (France); 1965–1969 K.V. Astachov (Russia); 1963–1971 R.G. Bates (USA); 1963–1967 H. Brusset (France); 1985– T. Cvitaš (Croatia); 1963 F. Daniels (USA); 1981–1987 E.T. Denisov (Russia); 1967–1975 M. Fayard (France); 1963–1965 J.I. Gerassimov (Russia); 1979–1987 K.H. Homann (Germany); 1963–1971 W. Jaenicke (Germany); 1967–1971 F. Jellinek (Netherlands); 1977–1985 N. Kallay (Croatia); 1973–1981 V. Kellö (Czechoslovakia); 1989– I.V. Khudyakov (Russia); 1985–1987 W.H. Kirchhoff (USA); 1971–1980 J. Koefoed (Denmark); 1979–1987 K. Kuchitsu (Japan); 1971–1981 D.R. Lide Jr (USA); 1963–1971 M.L. McGlashan (UK); 1983–1991 I.M. Mills (UK); 1963–1967 M. Milone (Italy); 1967–1973 M.A. Paul (USA); 1991– F. Pavese (Italy); 1963–1967 K.J. Pedersen (Denmark); 1967–1975 A. Perez-Masiá (Spain); 1987– M. Quack (Switzerland); 1971–1979 A. Schuyff (Netherlands); 1967–1970 L.G. Sillén (Sweden); 1989– H.L. Strauss (USA); 1963–1967 G. Waddington (USA); 1981–1985 D.D. Wagman (USA); 1971–1979 K.G. Weil (Germany); 1971–1977 D.H. Whiffen (UK); 1963–1967 E.H. Wiebenga (Netherlands).

Associate members

1983–1991 R.A. Alberty (USA); 1983–1987 I. Ansara (France); 1979–1991 E.R. Cohen (USA); 1979–1981 E.T. Denisov (Russia); 1987– G.H. Findenegg (Germany); 1987–1991 K.H. Homann (Germany); 1971–1973 W. Jaenicke (Germany); 1985–1989 N. Kallay (Croatia); 1987–1989 I.V. Khudyakov (Russia); 1987–1991 K. Kuchitsu (Japan); 1981–1983 D.R. Lide Jr (USA); 1971–1979 M.L. McGlashan (UK); 1991– I.M. Mills (UK); 1973–1981 M.A. Paul (USA); 1975–1983 A. Perez-Masiá (Spain); 1979–1987 A. Schuyff (Netherlands); 1963–1971 S. Seki (Japan); 1969–1977 J. Terrien (France); 1975–1979 L. Villena (Spain); 1967–1969 G. Waddington (USA); 1979–1983 K.G. Weil (Germany); 1977–1985 D.H. Whiffen (UK).

1
Physical quantities and units

1.1 PHYSICAL QUANTITIES AND QUANTITY CALCULUS

The value of a *physical quantity* can be expressed as the product of a *numerical value* and a *unit*:

$$\text{physical quantity} = \text{numerical value} \times \text{unit}$$

Neither the name of the physical quantity, nor the symbol used to denote it, should imply a particular choice of unit.

Physical quantities, numerical values, and units, may all be manipulated by the ordinary rules of algebra. Thus we may write, for example, for the wavelength λ of one of the yellow sodium lines:

$$\lambda = 5.896 \times 10^{-7}\ \text{m} = 589.6\ \text{nm} \qquad (1)$$

where m is the symbol for the unit of length called the metre (see chapter 3), nm is the symbol for the nanometre, and the units m and nm are related by

$$\text{nm} = 10^{-9}\ \text{m} \qquad (2)$$

The equivalence of the two expressions for λ in equation (1) follows at once when we treat the units by the rules of algebra and recognize the identity of nm and 10^{-9} m in equation (2). The wavelength may equally well be expressed in the form

$$\lambda/\text{m} = 5.896 \times 10^{-7} \qquad (3)$$

or

$$\lambda/\text{nm} = 589.6 \qquad (4)$$

In tabulating the numerical values of physical quantities, or labelling the axes of graphs, it is particularly convenient to use the quotient of a physical quantity and a unit in such a form that the values to be tabulated are pure numbers, as in equations (3) and (4).

Examples

T/K	10^3K/T	p/MPa	ln(p/MPa)
216.55	4.6179	0.5180	−0.6578
273.15	3.6610	3.4853	1.2486
304.19	3.2874	7.3815	1.9990

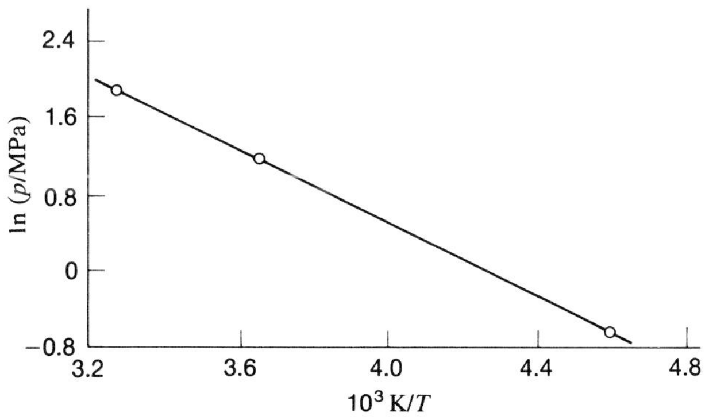

Algebraically equivalent forms may be used in place of 10^3K/T, such as kK/T or $10^3(T/\text{K})^{-1}$.

The method described here for handling physical quantities and their units is known as *quantity calculus*. It is recommended for use throughout science and technology. The use of quantity calculus does not imply any particular choice of units; indeed one of the advantages of quantity calculus is that it makes changes between units particularly easy to follow. Further examples of the use of quantity calculus are given in chapter 7, which is concerned with the problems of transforming from one set of units to another.

1.2 BASE PHYSICAL QUANTITIES AND DERIVED PHYSICAL QUANTITIES

By convention physical quantities are organized in a dimensional system built upon seven *base quantities*, each of which is regarded as having its own dimension. These base quantities and the symbols used to denote them are as follows:

Physical quantity	*Symbol for quantity*
length	l
mass	m
time	t
electric current	I
thermodynamic temperature	T
amount of substance	n
luminous intensity	I_v

All other physical quantities are called *derived quantities* and are regarded as having dimensions derived algebraically from the seven base quantities by multiplication and division.

Example dimension of (energy) = dimension of (mass $\times$ length^2 $\times$ time^{-2})

The physical quantity *amount of substance* or *chemical amount* is of special importance to chemists. Amount of substance is proportional to the number of specified elementary entities of that substance, the proportionality factor being the same for all substances; its reciprocal is the *Avogadro constant* (see sections 2.10, p.46, and 3.2, p.70, and chapter 5). The SI unit of amount of substance is the mole, defined in chapter 3 below. The physical quantity 'amount of substance' should no longer be called 'number of moles', just as the physical quantity 'mass' should not be called 'number of kilograms'. The name 'amount of substance' and 'chemical amount' may often be usefully abbreviated to the single word 'amount', particularly in such phrases as 'amount concentration' (p.42)[1], and 'amount of N_2' (see examples on p.46).

(1) The Clinical Chemistry Division of IUPAC recommends that 'amount-of-substance concentration' be abbreviated 'substance concentration'.

1.3 SYMBOLS FOR PHYSICAL QUANTITIES AND UNITS [5.a]

A clear distinction should be drawn between the names and symbols for physical quantities, and the names and symbols for units. Names and symbols for many physical quantities are given in chapter 2; the symbols given there are *recommendations*. If other symbols are used they should be clearly defined. Names and symbols for units are given in chapter 3; the symbols for units listed there are *mandatory*.

General rules for symbols for physical quantities

The symbol for a physical quantity should generally be a single letter of the Latin or Greek alphabet (see p.143)[1]. Capital and lower case letters may both be used. The letter should be printed in italic (sloping) type. When no italic font is available the distinction may be made by underlining symbols for physical quantities in accord with standard printers' practice. When necessary the symbol may be modified by subscripts and/or superscripts of specified meaning. Subscripts and superscripts that are themselves symbols for physical quantities or numbers should be printed in italic type; other subscripts and superscripts should be printed in roman (upright) type.

Examples		C_p	for heat capacity at constant pressure
		x_i	for mole fraction of the *i*th species
	but	C_B	for heat capacity of substance B
		E_k	for kinetic energy
		μ_r	for relative permeability
		$\Delta_\mathrm{r}H^\circ$	for standard reaction enthalpy
		V_m	for molar volume

The meaning of symbols for physical quantities may be further qualified by the use of one or more subscripts, or by information contained in round brackets.

Examples $\Delta_\mathrm{f}S^\circ(\mathrm{HgCl_2, cr, 25\,^\circ C}) = -154.3\ \mathrm{J\,K^{-1}\,mol^{-1}}$
$\mu_i = (\partial G/\partial n_i)_{T,p,n_{j\neq i}}$

Vectors and matrices may be printed in bold face italic type, e.g. $\boldsymbol{A}$, $\boldsymbol{a}$. Matrices and tensors are sometimes printed in bold face sans-serif type, e.g. $\boldsymbol{\mathsf{S}}, \boldsymbol{\mathsf{T}}$. Vectors may alternatively be characterized by an arrow, $\vec{A}$, $\vec{a}$ and second rank tensors by a double arrow, $\vec{\vec{S}}$, $\vec{\vec{T}}$.

General rules for symbols for units

Symbols for units should be printed in roman (upright) type. They should remain unaltered in the plural, and should not be followed by a full stop except at the end of a sentence.

Example $r = 10$ cm, not cm. or cms.

Symbols for units should be printed in lower case letters, unless they are derived from a personal name when they should begin with a capital letter. (An exception is the symbol for the litre which may be either L or l, i.e. either capital or lower case.)

---

(1) An exception is made for certain dimensionless quantities used in the study of transport processes for which the internationally agreed symbols consist of two letters (see section 2.15).

Example Reynolds number, *Re*

When such symbols appear as factors in a product, they should be separated from other symbols by a space, multiplication sign, or brackets.

Examples   m (metre), s (second), but J (joule), Hz (hertz)

Decimal multiples and submultiples of units may be indicated by the use of prefixes as defined in section 3.6 below.

Examples   nm (nanometre), kHz (kilohertz), Mg (megagram)

1.4 USE OF THE WORDS 'EXTENSIVE', 'INTENSIVE', 'SPECIFIC' AND 'MOLAR'

A quantity whose magnitude is additive for subsystems is called *extensive*; examples are mass m, volume V, Gibbs energy G. A quantity whose magnitude is independent of the extent of the system is called *intensive*; examples are temperature T, pressure p, chemical potential (partial molar Gibbs energy) μ.

The adjective *specific* before the name of an extensive quantity is often used to mean *divided by mass*. When the symbol for the extensive quantity is a capital letter, the symbol used for the specific quantity is often the corresponding lower case letter.

Examples volume, V
specific volume, $v = V/m = 1/\rho$ (where ρ is mass density)
heat capacity at constant pressure, C_p
specific heat capacity at constant pressure, $c_p = C_p/m$

ISO [5.a] recommends systematic naming of physical quantities derived by division with mass, volume, area and length by using the attributes massic, volumic, areic and lineic, respectively. In addition the Clinical Chemistry Division of IUPAC recommends the use of the attribute entitic for quantities derived by division with the number of entities [8]. Thus, for example, the specific volume is called massic volume and the surface charge density areic charge.

The adjective *molar* before the name of an extensive quantity generally means *divided by amount of substance*. The subscript m on the symbol for the extensive quantity denotes the corresponding molar quantity.

Examples	volume, V	molar volume, $V_m = V/n$ (p.41)
	enthalpy, H	molar enthalpy, $H_m = H/n$

It is sometimes convenient to divide all extensive quantities by amount of substance, so that all quantities become intensive; the subscript m may then be omitted if this convention is stated and there is no risk of ambiguity. (See also the symbols recommended for partial molar quantities in section 2.11, p.49, and 'Examples of the use of these symbols', p.51.)

There are a few cases where the adjective *molar* has a different meaning, namely *divided by amount-of-substance concentration*.

Examples absorption coefficient, a
molar absorption coefficient, $\varepsilon = a/c$ (p.32)
conductivity, κ
molar conductivity, $\Lambda = \kappa/c$ (p.60)

1.5 PRODUCTS AND QUOTIENTS OF PHYSICAL QUANTITIES AND UNITS

Products of physical quantities may be written in any of the ways

$$a\,b \quad \text{or} \quad ab \quad \text{or} \quad a \cdot b \quad \text{or} \quad a \times b$$

and similarly quotients may be written

$$a/b \quad \text{or} \quad \frac{a}{b} \quad \text{or} \quad ab^{-1}$$

Examples $F = ma$, $p = nRT/V$

Not more than one solidus (/) should be used in the same expression unless brackets are used to eliminate ambiguity.

Example $(a/b)/c$, but never $a/b/c$

In evaluating combinations of many factors, multiplication takes precedence over division in the sense that a/bc should be interpreted as $a/(bc)$ rather than $(a/b)c$; however, in complex expressions it is desirable to use brackets to eliminate any ambiguity.

Products and quotients of units may be written in a similar way, except that when a product of units is written without any multiplication sign one space should be left between the unit symbols.

Example $\mathrm{N = m\ kg\ s^{-2}}$, but not $\mathrm{mkgs^{-2}}$

2
Tables of physical quantities

The following tables contain the internationally recommended names and symbols for the physical quantities most likely to be used by chemists. Further quantities and symbols may be found in recommendations by IUPAP [4] and ISO [5].

Although authors are free to choose any symbols they wish for the quantities they discuss, provided that they define their notation and conform to the general rules indicated in chapter 1, it is clearly an aid to scientific communication if we all generally follow a standard notation. The symbols below have been chosen to conform with current usage and to minimize conflict so far as possible. Small variations from the recommended symbols may often be desirable in particular situations, perhaps by adding or modifying subscripts and/or superscripts, or by the alternative use of upper or lower case. Within a limited subject area it may also be possible to simplify notation, for example by omitting qualifying subscripts or superscripts, without introducing ambiguity. The notation adopted should in any case always be defined. Major deviations from the recommended symbols should be particularly carefully defined.

The tables are arranged by subject. The five columns in each table give the name of the quantity, the recommended symbol(s), a brief definition, the symbol for the coherent SI unit (without multiple or submultiple prefixes, see p.74), and footnote references. When two or more symbols are recommended, commas are used to separate symbols that are equally acceptable, and symbols of second choice are put in parentheses. A semicolon is used to separate symbols of slightly different quantities. The definitions are given primarily for identification purposes and are not necessarily complete; they should be regarded as useful relations rather than formal definitions. For dimensionless quantities a 1 is entered in the SI unit column. Further information is added in footnotes, and in text inserts between the tables, as appropriate.

2.1 SPACE AND TIME

The names and symbols recommended here are in agreement with those recommended by IUPAP [4] and ISO [5.b,c].

Name	*Symbol*	*Definition*	*SI unit*	*Notes*
cartesian space coordinates	x, y, z		m	
spherical polar coordinates	$r; \theta; \phi$		m, 1, 1	
cylindrical coordinates	$\rho; \theta; z$		m, 1, m	
generalized coordinate	q, q_i		(varies)	
position vector	$\boldsymbol{r}$	$\boldsymbol{r} = x\boldsymbol{i} + y\boldsymbol{j} + z\boldsymbol{k}$	m	
length	l		m	
special symbols:				
height	h			
breadth	b			
thickness	d, δ			
distance	d			
radius	r			
diameter	d			
path length	s			
length of arc	s			
area	A, A_s, S		m^2	1
volume	$V, (v)$		m^3	
plane angle	$\alpha, \beta, \gamma, \theta, \phi \ldots$	$\alpha = s/r$	rad, 1	2
solid angle	Ω, ω	$\Omega = A/r^2$	sr, 1	2
time	t		s	
period	T	$T = t/N$	s	
frequency	ν, f	$\nu = 1/T$	Hz	
angular frequency, circular frequency	ω	$\omega = 2\pi\nu$	$\mathrm{rad\ s^{-1}, s^{-1}}$	2, 3
characteristic time interval, relaxation time, time constant	τ, T	$\tau = \lvert \mathrm{d}t/\mathrm{d}\ln x \rvert$	s	
angular velocity	ω	$\omega = \mathrm{d}\phi/\mathrm{d}t$	$\mathrm{rad\ s^{-1}, s^{-1}}$	2, 4
velocity	$\boldsymbol{v}, \boldsymbol{u}, \boldsymbol{w}, \boldsymbol{c}, \dot{\boldsymbol{r}}$	$\boldsymbol{v} = \mathrm{d}\boldsymbol{r}/\mathrm{d}t$	$\mathrm{m\ s^{-1}}$	
speed	v, u, w, c	$v = \lvert \boldsymbol{v} \rvert$	$\mathrm{m\ s^{-1}}$	5
acceleration	$\boldsymbol{a}$	$\boldsymbol{a} = \mathrm{d}\boldsymbol{v}/\mathrm{d}t$	$\mathrm{m\ s^{-2}}$	6

(1) An infinitesimal area may be regarded as a vector d$\boldsymbol{A}$ perpendicular to the plane. The symbol A_s may be used when necessary to avoid confusion with A for Helmholtz energy.
(2) The units radian (rad) and steradian (sr), for plane angle and solid angle respectively, are described as 'SI supplementary units' [3]. Since they are of dimension 1 (i.e. dimensionless), they may be included if appropriate, or they may be omitted if clarity is not lost thereby, in expressions for derived SI units.
(3) The unit Hz is not to be used for angular frequency.
(4) Angular velocity can be treated as a vector.
(5) For the speeds of light and sound the symbol c is customary.
(6) For acceleration of free fall the symbol g is used.

2.2 CLASSICAL MECHANICS

The names and symbols recommended here are in agreement with those recommended by IUPAP [4] and ISO [5.d]. Additional quantities and symbols used in acoustics can be found in [4 and 5.h].

Name	*Symbol*	*Definition*	*SI unit*	*Notes*
mass	m		kg	
reduced mass	μ	$\mu = m_1 m_2/(m_1 + m_2)$	kg	
density, mass density	ρ	$\rho = m/V$	$\mathrm{kg\,m^{-3}}$	
relative density	d	$d = \rho/\rho^{\ominus}$	1	1
surface density	ρ_A, ρ_S	$\rho_A = m/A$	$\mathrm{kg\,m^{-2}}$	
specific volume	v	$v = V/m = 1/\rho$	$\mathrm{m^3\,kg^{-1}}$	
momentum	$\boldsymbol{p}$	$\boldsymbol{p} = m\boldsymbol{v}$	$\mathrm{kg\,m\,s^{-1}}$	
angular momentum, action	$\boldsymbol{L}$	$\boldsymbol{L} = \boldsymbol{r} \times \boldsymbol{p}$	J s	2
moment of inertia	I, J	$I = \sum m_i r_i{}^2$	$\mathrm{kg\,m^2}$	3
force	$\boldsymbol{F}$	$\boldsymbol{F} = \mathrm{d}\boldsymbol{p}/\mathrm{d}t = m\boldsymbol{a}$	N	
torque, moment of a force	$\boldsymbol{T}, (\boldsymbol{M})$	$\boldsymbol{T} = \boldsymbol{r} \times \boldsymbol{F}$	N m	
energy	E		J	
potential energy	E_p, V, Φ	$E_\mathrm{p} = -\int \boldsymbol{F} \cdot \mathrm{d}\boldsymbol{s}$	J	
kinetic energy	E_k, T, K	$E_\mathrm{k} = \frac{1}{2}mv^2$	J	
work	W, w	$W = \int \boldsymbol{F} \cdot \mathrm{d}\boldsymbol{s}$	J	
Lagrange function	L	$L(q, \dot{q}) = T(q, \dot{q}) - V(q)$	J	
Hamilton function	H	$H(q, p) = \sum p_i \dot{q}_i - L(q, \dot{q})$	J	
pressure	p, P	$p = F/A$	Pa, $\mathrm{N\,m^{-2}}$	
surface tension	γ, σ	$\gamma = \mathrm{d}W/\mathrm{d}A$	$\mathrm{N\,m^{-1}}$, $\mathrm{J\,m^{-2}}$	
weight	$G, (W, P)$	$G = mg$	N	
gravitational constant	G	$F = Gm_1 m_2/r^2$	$\mathrm{N\,m^2\,kg^{-2}}$	
normal stress	σ	$\sigma = F/A$	Pa	
shear stress	τ	$\tau = F/A$	Pa	
linear strain, relative elongation	ε, e	$\varepsilon = \Delta l/l$	1	
modulus of elasticity, Young's modulus	E	$E = \sigma/\varepsilon$	Pa	
shear strain	γ	$\gamma = \Delta x/d$	1	
shear modulus	G	$G = \tau/\gamma$	Pa	
volume strain, bulk strain	θ	$\theta = \Delta V/V_0$	1	
bulk modulus, compression modulus	K	$K = -V_0(\mathrm{d}p/\mathrm{d}V)$	Pa	

(1) Usually $\rho^{\ominus} = \rho(H_2O, 4\,°C)$.

(2) Other symbols are customary in atomic and molecular spectroscopy; see section 2.6.

(3) In general I is a tensor quantity: $I_{\alpha\alpha} = \Sigma m_i(\beta_i^2 + \gamma_i^2)$, and $I_{\alpha\beta} = -\Sigma m_i \alpha_i \beta_i$ if $\alpha \neq \beta$, where α, β, γ is a permutation of x, y, z. For a continuous distribution of mass the sums are replaced by integrals.

Name	*Symbol*	*Definition*	*SI unit*	*Notes*
viscosity, dynamic viscosity	η, μ	$\tau_{x,z} = \eta(\mathrm{d}v_x/\mathrm{d}z)$	Pa s	
fluidity	ϕ	$\phi = 1/\eta$	$\mathrm{m\,kg^{-1}\,s}$	
kinematic viscosity	ν	$\nu = \eta/\rho$	$\mathrm{m^2\,s^{-1}}$	
friction factor	μ, (f)	$F_{\mathrm{frict}} = \mu F_{\mathrm{norm}}$	1	
power	P	$P = \mathrm{d}W/\mathrm{d}t$	W	
sound energy flux	P, P_{a}	$P = \mathrm{d}E/\mathrm{d}t$	W	
acoustic factors,				
reflection	ρ	$\rho = P_{\mathrm{r}}/P_0$	1	4
absorption	α_{a}, (α)	$\alpha_{\mathrm{a}} = 1 - \rho$	1	5
transmission	τ	$\tau = P_{\mathrm{tr}}/P_0$	1	4
dissipation	δ	$\delta = \alpha_{\mathrm{a}} - \tau$	1	

(4) P_0 is the incident sound energy flux, P_{r} the reflected flux and P_{tr} the transmitted flux.
(5) This definition is special to acoustics and is different from the usage in radiation, where the absorption factor corresponds to the acoustic dissipation factor.

2.3 ELECTRICITY AND MAGNETISM

The names and symbols recommended here are in agreement with those recommended by IUPAP [4] and ISO [5.f].

Name	*Symbol*	*Definition*	*SI unit*	*Notes*
quantity of electricity, electric charge	Q		C	
charge density	ρ	$\rho = Q/V$	$\mathrm{C\,m^{-3}}$	
surface charge density	σ	$\sigma = Q/A$	$\mathrm{C\,m^{-2}}$	
electric potential	V, ϕ	$V = \mathrm{d}W/\mathrm{d}Q$	$\mathrm{V, J\,C^{-1}}$	
electric potential difference	$U, \Delta V, \Delta\phi$	$U = V_2 - V_1$	V	
electromotive force	E	$E = \int (\boldsymbol{F}/Q)\cdot \mathrm{d}\boldsymbol{s}$	V	
electric field strength	$\boldsymbol{E}$	$\boldsymbol{E} = \boldsymbol{F}/Q = -\nabla V$	$\mathrm{V\,m^{-1}}$	
electric flux	Ψ	$\Psi = \int \boldsymbol{D}\cdot \mathrm{d}\boldsymbol{A}$	C	1
electric displacement	$\boldsymbol{D}$	$\boldsymbol{D} = \varepsilon\boldsymbol{E}$	$\mathrm{C\,m^{-2}}$	
capacitance	C	$C = Q/U$	$\mathrm{F, C\,V^{-1}}$	
permittivity	ε	$\boldsymbol{D} = \varepsilon\boldsymbol{E}$	$\mathrm{F\,m^{-1}}$	
permittivity of vacuum	ε_0	$\varepsilon_0 = \mu_0^{-1} c_0^{-2}$	$\mathrm{F\,m^{-1}}$	
relative permittivity	ε_r	$\varepsilon_\mathrm{r} = \varepsilon/\varepsilon_0$	1	2
dielectric polarization (dipole moment per volume)	$\boldsymbol{P}$	$\boldsymbol{P} = \boldsymbol{D} - \varepsilon_0\boldsymbol{E}$	$\mathrm{C\,m^{-2}}$	
electric susceptibility	χ_e	$\chi_\mathrm{e} = \varepsilon_\mathrm{r} - 1$	1	
1st hyper-susceptibility	$\chi_\mathrm{e}^{(2)}$	$\chi_\mathrm{e}^{(2)} = \partial^2 P/\partial E^2$	$\mathrm{C\,m\,J^{-1}}$	3
2nd hyper-susceptibility	$\chi_\mathrm{e}^{(3)}$	$\chi_\mathrm{e}^{(3)} = \partial^3 P/\partial E^3$	$\mathrm{C^2\,m^2\,J^{-2}}$	3
electric dipole moment	$\boldsymbol{p}, \boldsymbol{\mu}$	$\boldsymbol{p} = \Sigma Q_i \boldsymbol{r}_i$	$\mathrm{C\,m}$	4
electric current	I, i	$I = \mathrm{d}Q/\mathrm{d}t$	A	
electric current density	$\boldsymbol{j}, \boldsymbol{J}$	$I = \int \boldsymbol{j}\cdot \mathrm{d}\boldsymbol{A}$	$\mathrm{A\,m^{-2}}$	1
magnetic flux density, magnetic induction	$\boldsymbol{B}$	$\boldsymbol{F} = Q\boldsymbol{v}\times\boldsymbol{B}$	T	5
magnetic flux	Φ	$\Phi = \int \boldsymbol{B}\cdot \mathrm{d}\boldsymbol{A}$	Wb	1
magnetic field strength	$\boldsymbol{H}$	$\boldsymbol{B} = \mu\boldsymbol{H}$	$\mathrm{A\,m^{-1}}$	

(1) $\mathrm{d}\boldsymbol{A}$ is a vector element of area.

(2) This quantity was formerly called dielectric constant.

(3) The hyper-susceptibilities are the coefficients of the non-linear terms in the expansion of the polarization $\boldsymbol{P}$ in powers of the electric field $\boldsymbol{E}$:

$$P = \varepsilon_0[\chi_\mathrm{e}^{(1)}E + (1/2)\chi_\mathrm{e}^{(2)}E^2 + (1/6)\chi_\mathrm{e}^{(3)}E^3 + \ldots]$$

where $\chi_\mathrm{e}^{(1)}$ is the usual electric susceptibility χ_e, equal to $\varepsilon_\mathrm{r} - 1$ in the absence of higher terms. In a medium that is anisotropic $\chi_\mathrm{e}^{(1)}$, $\chi_\mathrm{e}^{(2)}$ and $\chi_\mathrm{e}^{(3)}$ are tensors of rank 2, 3 and 4, respectively. For an isotropic medium (such as a liquid) or for a crystal with a centrosymmetric unit cell, $\chi_\mathrm{e}^{(2)}$ is zero by symmetry. These quantities characterize a dielectric medium in the same way that the polarizability and the hyper-polarizabilities characterize a molecule (see p.22).

(4) When a dipole is composed of two point charges Q and $-Q$ separated by a distance r, the direction of the dipole vector is taken to be from the negative to the positive charge. The opposite convention is sometimes used, but is to be discouraged. The dipole moment of an ion depends on the choice of the origin.

(5) This quantity is sometimes loosely called magnetic field.

Name	*Symbol*	*Definition*	*SI unit*	*Notes*
permeability	μ	$\boldsymbol{B} = \mu \boldsymbol{H}$	$\mathrm{N\,A^{-2}}$, $\mathrm{H\,m^{-1}}$	
permeability of vacuum	μ_0	$\mu_0 = 4\pi \times 10^{-7}\,\mathrm{H\,m^{-1}}$	$\mathrm{H\,m^{-1}}$	
relative permeability	μ_r	$\mu_\mathrm{r} = \mu/\mu_0$	1	
magnetization (magnetic dipole moment per volume)	$\boldsymbol{M}$	$\boldsymbol{M} = \boldsymbol{B}/\mu_0 - \boldsymbol{H}$	$\mathrm{A\,m^{-1}}$	
magnetic susceptibility	χ, κ, (χ_m)	$\chi = \mu_\mathrm{r} - 1$	1	6
molar magnetic susceptibility	χ_m	$\chi_\mathrm{m} = V_\mathrm{m}\chi$	$\mathrm{m^3\,mol^{-1}}$	
magnetic dipole moment	$\boldsymbol{m}$, $\boldsymbol{\mu}$	$E_\mathrm{p} = -\boldsymbol{m}\cdot\boldsymbol{B}$	$\mathrm{A\,m^2}$, $\mathrm{J\,T^{-1}}$	
electric resistance	R	$R = U/I$	Ω	7
conductance	G	$G = 1/R$	S	7
loss angle	δ	$\delta = \phi_I - \phi_U$	1, rad	8
reactance	X	$X = (U/I)\sin\delta$	Ω	
impedance, (complex impedance)	Z	$Z = R + \mathrm{i}X$	Ω	
admittance, (complex admittance)	Y	$Y = 1/Z$	S	
susceptance	B	$Y = G + \mathrm{i}B$	S	
resistivity	ρ	$\rho = E/j$	$\Omega\,\mathrm{m}$	9
conductivity	κ, γ, σ	$\kappa = 1/\rho$	$\mathrm{S\,m^{-1}}$	9
self-inductance	L	$E = -L(\mathrm{d}I/\mathrm{d}t)$	H	
mutual inductance	M, L_{12}	$E_1 = L_{12}(\mathrm{d}I_2/\mathrm{d}t)$	H	
magnetic vector potential	$\boldsymbol{A}$	$\boldsymbol{B} = \nabla \times \boldsymbol{A}$	$\mathrm{Wb\,m^{-1}}$	
Poynting vector	$\boldsymbol{S}$	$\boldsymbol{S} = \boldsymbol{E} \times \boldsymbol{H}$	$\mathrm{W\,m^{-2}}$	10

(6) The symbol χ_m is sometimes used for magnetic susceptibility, but it should be reserved for molar magnetic susceptibility.

(7) In a material with reactance $R = (U/I)\cos\delta$, and $G = R/(R^2 + X^2)$.

(8) ϕ_I and ϕ_U are the phases of current and potential difference.

(9) These quantities are tensors in anisotropic materials.

(10) This quantity is also called the Poynting–Umov vector.

2.4 QUANTUM MECHANICS AND QUANTUM CHEMISTRY

The names and symbols for quantities used in quantum mechanics and recommended here are in agreement with those recommended by IUPAP [4]. The names and symbols for quantities used mainly in the field of quantum chemistry have been chosen on the basis of the current practice in the field.

Name	*Symbol*	*Definition*	*SI unit*	*Notes*
momentum operator	$\hat{\boldsymbol{p}}$	$\hat{\boldsymbol{p}} = -\mathrm{i}\hbar\nabla$	$\mathrm{J\,s\,m^{-1}}$	1
kinetic energy operator	$\hat{T}$	$\hat{T} = -(\hbar^2/2m)\nabla^2$	J	1
hamiltonian operator	$\hat{H}$	$\hat{H} = \hat{T} + \hat{V}$	J	1
wavefunction, state function	Ψ, ψ, ϕ	$\hat{H}\psi = E\psi$	$(\mathrm{m^{-3/2}})$	2, 3
hydrogen-like wavefunction	$\psi_{nlm}(r, \theta, \phi)$	$\psi_{nlm} = R_{nl}(r)\, Y_{lm}(\theta, \phi)$	$(\mathrm{m^{-3/2}})$	3
spherical harmonic function	$Y_{lm}(\theta, \phi)$	$Y_{lm} = N_{l,\lvert m\rvert} P_l^{\lvert m\rvert}(\cos\theta)\mathrm{e}^{\mathrm{i}m\phi}$	1	4
probability density	P	$P = \psi^*\psi$	$(\mathrm{m^{-3}})$	3, 5
charge density of electrons	ρ	$\rho = -eP$	$(\mathrm{C\,m^{-3}})$	3, 5, 6
probability current density, probability flux	$\boldsymbol{S}$	$\boldsymbol{S} = -(\mathrm{i}\hbar/2m) \times(\psi^*\nabla\psi - \psi\nabla\psi^*)$	$(\mathrm{m^{-2}\,s^{-1}})$	3
electric current density of electrons	$\boldsymbol{j}$	$\boldsymbol{j} = -e\boldsymbol{S}$	$(\mathrm{A\,m^{-2}})$	3, 6
integration element	$\mathrm{d}\tau$	$\mathrm{d}\tau = \mathrm{d}x\,\mathrm{d}y\,\mathrm{d}z$, etc.	(varies)	
matrix element of operator $\hat{A}$	A_{ij}, $\langle i \lvert A \rvert j\rangle$	$A_{ij} = \int\psi_i^*\hat{A}\psi_j\mathrm{d}\tau$	(varies)	7
expectation value of operator $\hat{A}$	$\langle A\rangle$, $\bar{A}$	$\langle A\rangle = \int\psi^*\hat{A}\psi\,\mathrm{d}\tau$	(varies)	7
hermitian conjugate of $\hat{A}$	$\hat{A}^\dagger$	$(A^\dagger)_{ij} = (A_{ji})^*$	(varies)	7
commutator of $\hat{A}$ and $\hat{B}$	$[\hat{A}, \hat{B}]$, $[\hat{A}, \hat{B}]_-$	$[\hat{A}, \hat{B}] = \hat{A}\hat{B} - \hat{B}\hat{A}$	(varies)	8
anticommutator of $\hat{A}$ and $\hat{B}$	$[\hat{A}, \hat{B}]_+$	$[\hat{A}, \hat{B}]_+ = \hat{A}\hat{B} + \hat{B}\hat{A}$	(varies)	8

(1) The 'hat' (or circumflex), ^, is used to distinguish an operator from an algebraïc quantity. ∇ denotes the nabla operator (see section 4.2, p.85).
(2) Capital and lower case psi are often used for the time-dependent function $\Psi(x, t)$ and the amplitude function $\psi(x)$ respectively. Thus for a stationary state $\Psi(x, t) = \psi(x)\exp(-\mathrm{i}Et/\hbar)$.
(3) For the normalized wavefunction of a single particle in three-dimensional space the appropriate SI unit is given in parentheses. Results in quantum chemistry, however, are often expressed in terms of atomic units (see section 3.8, p.76; section 7.3, p.120; and reference [9]). If distances, energies, angular momenta, charges and masses are all expressed as dimensionless ratios r/a_0, E/E_h, $L/\hbar$, Q/e, and m/m_e respectively, then all quantities are dimensionless.
(4) $P_l^{\lvert m\rvert}$ denotes the associated Legendre function of degree l and order $\lvert m\rvert$. $N_{l,\lvert m\rvert}$ is a normalization constant.
(5) ψ^* is the complex conjugate of ψ. For an antisymmetrized n electron wavefunction $\Psi(r_1, \ldots, r_n)$, the total probability density of electrons is $\int_2 \ldots \int_n \Psi^*\Psi\,\mathrm{d}\tau_2 \ldots \mathrm{d}\tau_n$, where the integration extends over the coordinates of all electrons but one.
(6) $-e$ is the charge of an electron.
(7) The unit is the same as for the physical quantity A that the operator represents.
(8) The unit is the same as for the product of the physical quantities A and B.

Name	*Symbol*	*Definition*	*SI unit*	*Notes*
angular momentum operators	—see p.26			
spin wavefunction	α; β		1	9

***Hückel molecular orbital theory* (HMO):**

Name	*Symbol*	*Definition*	*SI unit*	*Notes*
atomic orbital basis function	χ_r		$m^{-3/2}$	3
molecular orbital	ϕ_i	$\phi_i = \sum_r \chi_r c_{ri}$	$m^{-3/2}$	3, 10
coulomb integral	H_{rr}, α	$H_{rr} = \int \chi_r^* \hat{H} \chi_r \, d\tau$	J	3, 10, 11
resonance integral	H_{rs}, β	$H_{rs} = \int \chi_r^* \hat{H} \chi_s \, d\tau$	J	3, 10
energy parameter	x	$x = (\alpha - E)/\beta$	1	12
overlap integral	S_{rs}	$S_{rs} = \int \chi_r^* \chi_s \, d\tau$	1	10
charge density	q_r	$q_r = \sum_i^{occ} c_{ri}^2$	1	13
bond order	p_{rs}	$p_{rs} = \sum_i^{occ} c_{ri} c_{si}$	1	13

(9) The spin wavefunctions of a single electron, α and β, are defined by the matrix elements of the z component of the spin angular momentum, $\hat{s}_z$, by the relations $\langle \alpha | \hat{s}_z | \alpha \rangle = +\frac{1}{2}$, $\langle \beta | \hat{s}_z | \beta \rangle = -\frac{1}{2}$, $\langle \alpha | \hat{s}_z | \beta \rangle = \langle \beta | \hat{s}_z | \alpha \rangle = 0$. The total electron spin wavefunctions of an atom with many electrons are denoted by Greek letters α, β, γ, etc. according to the value of $\sum m_S$, starting from the highest down to the lowest.

(10) $\hat{H}$ is an effective hamiltonian for a single electron, i and j label the molecular orbitals, and r and s label the atomic orbitals. In Hückel MO theory H_{rs} is taken to be non-zero only for bonded pairs of atoms r and s, and all S_{rs} are assumed to be zero for $r \neq s$.

(11) Note that the name 'coulomb integral' has a different meaning in HMO theory (where it refers to the energy of the orbital χ_r in the field of the nuclei) to Hartree–Fock theory discussed below (where it refers to a two-electron repulsion integral).

(12) In the simplest application of Hückel theory to the π electrons of planar conjugated hydrocarbons, α is taken to be the same for all C atoms, and β to be the same for all bonded pairs of C atoms; it is then customary to write the Hückel secular determinant in terms of the dimensionless parameter x.

(13) $-eq_r$ is the charge on atom r, and p_{rs} is the bond order between atoms r and s. The sum goes over all occupied molecular spin-orbitals.

***Ab initio Hartree–Fock self-consistent field theory* (ab initio SCF)**

Results in quantum chemistry are often expressed in atomic units (see p.76 and p.120). In the remaining tables of this section all lengths, energies, masses, charges and angular momenta are expressed as dimensionless ratios to the corresponding atomic units, a_0, E_h, m_e, e and $\hbar$ respectively. Thus all quantities become dimensionless, and the SI unit column is omitted.

Name	*Symbol*	*Definition*	*Notes*
molecular orbital	$\phi_i(\mu)$		14
molecular spin-orbital	$\phi_i(\mu)\alpha(\mu)$; $\phi_i(\mu)\beta(\mu)$		14
total wavefunction	Ψ	$\Psi = (N!)^{-\frac{1}{2}} \lVert \phi_i(\mu) \rVert$	14, 15
core hamiltonian of a single electron	$\hat{H}_\mu^{\text{core}}$	$\hat{H}_\mu = -\frac{1}{2}\nabla_\mu^2 - \sum_{\text{A}} Z_{\text{A}}/r_{\mu\text{A}}$	14, 16
one-electron integrals: expectation value of the core hamiltonian	H_{ii}	$H_{ii} = \int \phi_i^*(1)\hat{H}_1^{\text{core}}\phi_i(1)\,\mathrm{d}\tau_1$	14, 16
two-electron repulsion integrals:			
coulomb integral	J_{ij}	$J_{ij} = \iint \phi_i^*(1)\phi_j^*(2)\frac{1}{r_{12}}\phi_i(1)\phi_j(2)\,\mathrm{d}\tau_1\,\mathrm{d}\tau_2$	14, 17
exchange integral	K_{ij}	$K_{ij} = \iint \phi_i^*(1)\phi_j^*(2)\frac{1}{r_{12}}\phi_j(1)\phi_i(2)\,\mathrm{d}\tau_1\,\mathrm{d}\tau_2$	14, 17
one-electron orbital energy	ε_i	$\varepsilon_i = H_{ii} + \sum_j (2J_{ij} - K_{ij})$	14, 18
total electronic energy	E	$E = 2\sum_i H_{ii} + \sum_i\sum_j (2J_{ij} - K_{ij})$ $= \sum_i (\varepsilon_i + H_{ii})$	14, 18, 19
coulomb operator	$\hat{J}_i$	$\hat{J}_i\phi_j(2) = \langle \phi_i(1) \left\lvert \frac{1}{r_{12}} \right\rvert \phi_i(1) \rangle \phi_j(2)$	14
exchange operator	$\hat{K}_i$	$\hat{K}_i\phi_j(2) = \langle \phi_i(1) \left\lvert \frac{1}{r_{12}} \right\rvert \phi_j(1) \rangle \phi_i(2)$	14
Fock operator	$\hat{F}$	$\hat{F} = \hat{H}^{\text{core}} + \sum_i (2\hat{J}_i - \hat{K}_i)$	14, 20

(14) The indices i and j label the molecular orbitals, and either μ or the numerals 1 and 2 label the electron coordinates.
(15) The double bars denote an antisymmetrized product of the occupied molecular spin-orbitals $\phi_i\alpha$ and $\phi_i\beta$ (sometimes denoted ϕ_i and $\bar{\phi}_i$); for a closed shell system Ψ would be a normalized Slater determinant. $(N!)^{-\frac{1}{2}}$ is the normalization constant.
(16) Z_{A} is the charge number (atomic number) of nucleus A, and $r_{\mu\text{A}}$ is the distance of electron μ from nucleus A. H_{ii} is the energy of an electron in orbital ϕ_i in the field of the core.
(17) The inter-electron repulsion integrals are often written in a contracted form as follows: $J_{ij} = (ii^* \mid jj^*)$, and $K_{ij} = (i^*j \mid ij^*)$. It is conventionally understood that the first two indices within the bracket refer to the orbitals involving electron 1, and the second two indices to the orbitals involving electron 2. In general the functions are real and the stars * are omitted.
(18) These relations apply to closed shell systems only, and the sums extend over the occupied molecular orbitals.
(19) The sum over j includes the term with $j = i$, for which $J_{ii} = K_{ii}$, so that this term in the sum simplifies to give $2J_{ii} - K_{ii} = J_{ii}$.
(20) The Hartree–Fock equations read $(\hat{F} - \varepsilon_j)\phi_j = 0$. Note that the definition of the Fock operator involves all of its eigenfuctions ϕ_i through the coulomb and exchange operators, $\hat{J}_i$ and $\hat{K}_i$.

Hartree–Fock–Roothaan SCF theory, using molecular orbitals expanded as linear combinations of atomic orbital basis functions (LCAO–MO theory)

Name	*Symbol*	*Definition*	*Notes*
atomic orbital basis function	χ_r		21
molecular orbital	ϕ_i	$\phi_i = \sum_r \chi_r c_{ri}$	
overlap matrix element	S_{rs}	$S_{rs} = \int \chi_r^* \chi_s \, \mathrm{d}\tau, \; \sum_{r,s} c_{ri}^* S_{rs} c_{sj} = \delta_{ij}$	
density matrix element	P_{rs}	$P_{rs} = 2 \sum_i^{\mathrm{occ}} c_{ri} c_{si}^*$	22
integrals over the basis functions:			
one-electron integrals	H_{rs}	$H_{rs} = \int \chi_r^*(1) \hat{H}_1^{\mathrm{core}} \chi_s(1) \, \mathrm{d}\tau_1$	
two-electron integrals	$(rs \vert tu)$	$(rs \vert tu) = \iint \chi_r(1) \chi_s(1) \frac{1}{r_{12}} \chi_t(2) \chi_u(2) \, \mathrm{d}\tau_1 \, \mathrm{d}\tau_2$	23, 24
total electronic energy	E	$E = \sum_r \sum_s P_{rs} H_{rs} + \frac{1}{2} \sum_r \sum_s \sum_t \sum_u P_{rs} P_{tu} [(rs \vert tu) - \frac{1}{2}(rt \vert su)]$	24
matrix element of the Fock operator	F_{rs}	$F_{rs} = H_{rs} + \sum_t \sum_u P_{tu} [(rs \vert tu) - \frac{1}{2}(rt \vert su)]$	25

(21) The indices r and s label the basis functions. In numerical computations the basis functions are either taken as Slater-type orbitals (STOs) or as gaussian type orbitals (GTOs). An STO basis function in spherical polar coordinates has the general form $\chi(r, \theta, \phi) = N r^{n-1} \exp(-\zeta_{nl} r) Y_{lm}(\theta, \phi)$, where ζ_{nl} is a shielding parameter representing the effective charge in the state with quantum numbers n and l. GTO functions are usually expressed in cartesian coordinates, in the form $\chi(x, y, z) = N x^a y^b z^c \exp(-\alpha r^2)$. Often a linear combination of two or three such functions with varying exponents α is used, in such a way as to model an STO. N denotes a normalization constant.

(22) The sum goes over all occupied molecular orbitals.

(23) The contracted notation for two-electron integrals over the basis functions, $(rs|tu)$, is based on the same convention outlined in note (17).

(24) Here the quantities are expressed in terms of integrals over the basis functions. The matrix elements H_{ii}, J_{ij} and K_{ij} may be similarly expressed in terms of integrals over the basis functions according to the following equations:

$$H_{ii} = \sum_r \sum_s c_{ri}^* c_{si} H_{rs}$$

$$J_{ij} = \sum_r \sum_s \sum_t \sum_u c_{ri}^* c_{sj}^* c_{ti} c_{uj} (rt|su)$$

$$K_{ij} = \sum_r \sum_s \sum_t \sum_u c_{ri}^* c_{sj}^* c_{ti} c_{uj} (rs|tu)$$

(25) The Hartree–Fock–Roothaan SCF equations, expressed in terms of the matrix elements of the Fock operator F_{rs}, and the overlap matrix elements S_{rs}, take the form:

$$\sum_s (F_{rs} - \varepsilon_i S_{rs}) c_{si} = 0$$

2.5 ATOMS AND MOLECULES

The names and symbols recommended here are in agreement with those recommended by IUPAP [4] and ISO [5.j]. Additional quantities and symbols used in atomic, nuclear and plasma physics can be found in [4 and 5.k].

Name	*Symbol*	*Definition*	*SI unit*	*Notes*
nucleon number, mass number	A		1	
proton number, atomic number	Z		1	
neutron number	N	$N = A - Z$	1	
electron rest mass	m_e		kg	1, 2
mass of atom, atomic mass	m_a, m		kg	
atomic mass constant	m_u	$m_u = m_a(^{12}C)/12$	kg	1, 3
mass excess	Δ	$\Delta = m_a - Am_u$	kg	
elementary charge, proton charge	e		C	2
Planck constant	h		J s	
Planck constant/2π	$\hbar$	$\hbar = h/2\pi$	J s	2
Bohr radius	a_0	$a_0 = 4\pi\varepsilon_0\hbar^2/m_e e^2$	m	2
Hartree energy	E_h	$E_h = \hbar^2/m_e a_0{}^2$	J	2
Rydberg constant	R_∞	$R_\infty = E_h/2hc$	m^{-1}	
fine structure constant	α	$\alpha = e^2/4\pi\varepsilon_0\hbar c$	1	
ionization energy	E_i		J	
electron affinity	E_{ea}		J	
electronegativity	χ	$\chi = \frac{1}{2}(E_i + E_{ea})$	J	4
dissociation energy	E_d, D		J	
from the ground state	D_0		J	5
from the potential minimum	D_e		J	5

(1) Analogous symbols are used for other particles with subscripts: p for proton, n for neutron, a for atom, N for nucleus, etc.
(2) This quantity is also used as an atomic unit; see sections 3.8 and 7.3.
(3) m_u is equal to the unified atomic mass unit, with symbol u, i.e. $m_u = 1$ u (see section 3.7). In biochemistry the name dalton, with symbol Da, is used for the unified atomic mass unit, although the name and symbol have not been accepted by CGPM.
(4) The concept of electronegativity was introduced by L. Pauling as the power of an atom in a molecule to attract electrons to itself. There are several ways of defining this quanity [49]. The one given in the table has a clear physical meaning of energy and is due to R.S. Mulliken. The most frequently used scale, due to Pauling, is based on bond dissociation energies in eV and it is relative in the sense that the values are dimensionless and that only electronegativity differences are defined. For atoms A and B

$$\chi_{r,A} - \chi_{r,B} = (eV)^{-1/2}\sqrt{E_d(AB) - [E_d(AA) + E_d(BB)]}$$

where χ_r denotes the Pauling relative electronegativity. The scale is chosen so as to make the relative electronegativity of hydrogen $\chi_{r,H} = 2.1$. There is a difficulty in choosing the sign of the square root, which determines the sign of $\chi_{r,A} - \chi_{r,B}$. Pauling made this choice intuitively.
(5) The symbols D_0 and D_e are mainly used for diatomic dissociation energies.

Name	*Symbol*	*Definition*	*SI unit*	*Notes*
principal quantum number (H atom)	n	$E = -hcR/n^2$	1	
angular momentum quantum numbers	see under Spectroscopy, section 2.6			
magnetic dipole moment of a molecule	$\boldsymbol{m}, \boldsymbol{\mu}$	$E_p = -\boldsymbol{m} \cdot \boldsymbol{B}$	$J\,T^{-1}$	6
magnetizability of a molecule	ξ	$\boldsymbol{m} = \xi \boldsymbol{B}$	$J\,T^{-2}$	
Bohr magneton	μ_B	$\mu_B = e\hbar/2m_e$	$J\,T^{-1}$	
nuclear magneton	μ_N	$\mu_N = (m_e/m_p)\mu_B$	$J\,T^{-1}$	
magnetogyric ratio (gyromagnetic ratio)	γ	$\gamma = \mu/L$	$s^{-1}\,T^{-1}$	7
g-factor	g	$g = 2\mu/\mu_B$	1	
nuclear g-factor	g_N	$g_N = \mu/I\mu_N$	1	
Larmor angular frequency	ω_L	$\omega_L = (e/2m)B$	s^{-1}	8
Larmor frequency	ν_L	$\nu_L = \omega_L/2\pi$	Hz	
relaxation time,				
longitudinal	T_1		s	9
transverse	T_2		s	9
electric dipole moment of a molecule	$\boldsymbol{p}, \boldsymbol{\mu}$	$E_p = -\boldsymbol{p} \cdot \boldsymbol{E}$	C m	10
quadrupole moment of a molecule	$\boldsymbol{Q}; \boldsymbol{\Theta}$	$E_p = \frac{1}{2}\boldsymbol{Q}: \boldsymbol{V}'' = \frac{1}{3}\boldsymbol{\Theta}: \boldsymbol{V}''$	$C\,m^2$	11
quadrupole moment of a nucleus	eQ	$eQ = 2\langle \Theta_{zz} \rangle$	$C\,m^2$	12

(6) Magnetic moments of specific particles may be denoted by subscripts, e.g. μ_e, μ_p, μ_n for an electron, a proton, and a neutron. Tabulated values usually refer to the maximum expectation value of the z component. Values for stable nuclei are given in table 6.3.

(7) μ is the magnetic moment, L the angular momentum.

(8) This quantity is commonly called Larmor circular frequency.

(9) These quantities are used in the context of saturation effects in spectroscopy, particularly spin-resonance spectroscopy (see p.25–26).

(10) See footnote 7 on p.24.

(11) The quadrupole moment of a molecule may be represented either by the tensor $\boldsymbol{Q}$, defined by an integral over the charge density ρ:

$$Q_{\alpha\beta} = \int r_\alpha r_\beta \rho \, dV$$

where α and β denote x, y or z, or by the traceless tensor $\boldsymbol{\Theta}$ defined by

$$\Theta_{\alpha\beta} = (1/2)\int (3r_\alpha r_\beta - \delta_{\alpha\beta} r^2)\rho \, dV$$
$$= (1/2)[3Q_{\alpha\beta} - \delta_{\alpha\beta}(Q_{xx} + Q_{yy} + Q_{zz})]$$

$\boldsymbol{V}''$ is the second derivative of the electric potential:

$$V_{\alpha\beta}'' = -q_{\alpha\beta} = \partial^2 V/\partial\alpha\partial\beta$$

(12) Nuclear quadrupole moments are conventionally defined in a different way from molecular quadrupole moments. Q is an area and e is the elementary charge. eQ is taken to be the maximum expectation value of the zz tensor element. The values of Q for some nuclei are listed in table 6.3.

Name	*Symbol*	*Definition*	*SI unit*	*Notes*
electric field gradient tensor	$\boldsymbol{q}$	$q_{\alpha\beta} = -\partial^2 V/\partial\alpha\,\partial\beta$	$\mathrm{V\ m^{-2}}$	
quadrupole interaction energy tensor	$\boldsymbol{\chi}$	$\chi_{\alpha\beta} = eQq_{\alpha\beta}$	J	13
electric polarizability of a molecule	$\boldsymbol{\alpha}$	$\alpha_{ab} = \partial p_a/\partial E_b$	$\mathrm{C^2\ m^2\ J^{-1}}$	14
1st hyper-polarizability	$\boldsymbol{\beta}$	$\beta_{abc} = \partial^2 p_a/\partial E_b \partial E_c$	$\mathrm{C^3\ m^3\ J^{-2}}$	14
2nd hyper-polarizability	γ	$\gamma_{abcd} = \partial^3 p_a/\partial E_b \partial E_c \partial E_d$	$\mathrm{C^4\ m^4\ J^{-3}}$	14
activity (of a radioactive substance)	A	$A = -\mathrm{d}N_\mathrm{B}/\mathrm{d}t$	Bq	15
decay (rate) constant, disintegration (rate) constant	λ, k	$A = \lambda N_\mathrm{B}$	$\mathrm{s^{-1}}$	15
half life	$t_{\frac{1}{2}}$, $T_{\frac{1}{2}}$	$N_\mathrm{B}(t_{\frac{1}{2}}) = N_\mathrm{B}(0)/2$	s	15, 16
mean life	τ	$\tau = 1/\lambda$	s	16
level width	Γ	$\Gamma = \hbar/\tau$	J	
disintegration energy	Q		J	
cross section (of a nuclear reaction)	σ		$\mathrm{m^2}$	

(13) The nuclear quadrupole interaction energy tensor $\boldsymbol{\chi}$ is usually quoted in MHz, corresponding to the value of eQq/h, although the h is usually omitted.

(14) The polarizability $\boldsymbol{\alpha}$ and the hyper-polarizabilities β, γ, ... are the coefficients in the expansion of the dipole moment $\boldsymbol{p}$ in powers of the electric field $\boldsymbol{E}$ according to the equation:

$$\boldsymbol{p} = \boldsymbol{p}^{(0)} + \boldsymbol{\alpha E} + (1/2)\boldsymbol{\beta E}^2 + (1/6)\gamma \boldsymbol{E}^3 + \ldots$$

where $\boldsymbol{\alpha}$, $\boldsymbol{\beta}$ and γ are tensors of rank 2, 3 and 4, respectively. The components of these tensors are distinguished by the subscript indices abc ... as indicated in the definitions, the first index a always denoting the component of p, and the later indices the components of the electric field. The polarizability and the hyper-polarizabilities exhibit symmetry properties. Thus $\boldsymbol{\alpha}$ is usually a symmetric tensor, and all components of $\boldsymbol{\beta}$ are zero for a molecule with a centre of symmetry, etc. Values of the polarizabilities are often quoted in atomic units (see p.76), in the form $\alpha/4\pi\varepsilon_0$ in units $a_0{}^3$, $\beta/(4\pi\varepsilon_0)^2$ in units of $a_0{}^5 e^{-1}$, and $\gamma/(4\pi\varepsilon_0)^3$ in units of $a_0{}^7 e^{-2}$, etc.

(15) N_B is the number of radioactive atoms B.

(16) Half lives and mean lives are often given in years (a), see p.111. $t_{\frac{1}{2}} = \tau \ln 2$ for exponential decays.

2.6 SPECTROSCOPY

This section has been considerably extended compared with the first editions of the Manual [1.a–c] and with the corresponding section in the IUPAP document [4]. It is based on the recommendations of the ICSU Joint Commission for Spectroscopy [50, 51] and current practice in the field which is well represented in the books by Herzberg [52]. The IUPAC Commission on Molecular Structure and Spectroscopy has also published various recommendations which have been taken into account [10–16].

Name	*Symbol*	*Definition*	*SI unit*	*Notes*
total term	T	$T = E_{tot}/hc$	m^{-1}	1, 2
transition wavenumber	$\tilde{\nu}$, (ν)	$\tilde{\nu} = T' - T''$	m^{-1}	1
transition frequency	ν	$\nu = (E' - E'')/h$	Hz	
electronic term	T_e	$T_e = E_e/hc$	m^{-1}	1, 2
vibrational term	G	$G = E_{vib}/hc$	m^{-1}	1, 2
rotational term	F	$F = E_{rot}/hc$	m^{-1}	1, 2
spin–orbit coupling constant	A	$T_{s.o.} = A\langle \hat{\boldsymbol{L}} \cdot \hat{\boldsymbol{S}} \rangle$	m^{-1}	1
principal moments of inertia	I_A; I_B; I_C	$I_A \leqslant I_B \leqslant I_C$	$kg\ m^2$	
rotational constants,				
in wavenumber	$\tilde{A}$; $\tilde{B}$; $\tilde{C}$	$\tilde{A} = h/8\pi^2 c I_A$	m^{-1}	1, 2
in frequency	A; B; C	$A = h/8\pi^2 I_A$	Hz	
inertial defect	Δ	$\Delta = I_C - I_A - I_B$	$kg\ m^2$	
asymmetry parameter	κ	$\kappa = \dfrac{2B - A - C}{A - C}$	1	3
centrifugal distortion constants,				
S reduction	D_J; D_{JK}; D_K; d_1; d_2		m^{-1}	4
A reduction	Δ_J; Δ_{JK}; Δ_K; δ_J; δ_K		m^{-1}	4
harmonic vibration wavenumber	ω_e; ω_r		m^{-1}	5
vibrational anharmonicity constant	$\omega_e x_e$; x_{rs}; $g_{tt'}$		m^{-1}	5
vibrational quantum numbers	v_r; l_t		1	5

(1) In spectroscopy the unit cm^{-1} is almost always used for wavenumber, and term values and wavenumbers always refer to the reciprocal wavelength of the equivalent radiation in vacuum. The symbol c in the definition E/hc refers to the speed of light in vacuum.

(2) Term values and rotational constants are sometimes defined in wavenumber units (e.g. $T = E/hc$), and sometimes in frequency units (e.g. $T = E/h$). When the symbol is otherwise the same, it is convenient to distinguish wavenumber quantities with a tilde (e.g. $\tilde{\nu}$, $\tilde{T}$, $\tilde{A}$, $\tilde{B}$, $\tilde{C}$ for quantities defined in wavenumber units), although this is not a universal practice.

(3) The Wang asymmetry parameters are also used: for a near prolate top $b_p = (C - B)/(2A - B - C)$, and for a near oblate top $b_o = (A - B)/(2C - A - B)$.

(4) S and A stand for the symmetric and asymmetric reductions of the rotational hamiltonian respectively; see [53] for more details on the various possible representations of the centrifugal distortion constants.

(5) For a diatomic: $G(v) = \omega_e(v + \frac{1}{2}) - \omega_e x_e(v + \frac{1}{2})^2 + \ldots$. For a polyatomic molecule the $3N - 6$ vibrational modes ($3N - 5$ if linear) are labelled by the indices $r, s, t, \ldots$, or $i, j, k, \ldots$. The index r is usually assigned in descending wavenumber order, symmetry species by symmetry species. The index t is kept for degenerate modes. The vibrational term formula is

$$G(v) = \sum_r \omega_r(v_r + d_r/2) + \sum_{r \leqslant s} x_{rs}(v_r + d_r/2)(v_s + d_s/2) + \sum_{t \leqslant t'} g_{tt'} l_t l_{t'} + \cdots$$

Name	*Symbol*	*Definition*	*SI unit*	*Notes*
Coriolis zeta constant	ζ_{rs}^{α}		1	
angular momentum quantum numbers	see additional information below			
degeneracy, statistical weight	g, d, β		1	6
electric dipole moment of a molecule	$\boldsymbol{p}, \boldsymbol{\mu}$	$E_p = -\boldsymbol{p}\cdot\boldsymbol{E}$	C m	7
transition dipole moment of a molecule	$\boldsymbol{M}, \boldsymbol{R}$	$\boldsymbol{M} = \int \psi'^* \boldsymbol{p} \psi'' \mathrm{d}\tau$	C m	7, 8
interatomic distances,				9, 10
equilibrium	r_e		m	
zero-point average	r_z		m	
ground state	r_0		m	
substitution structure	r_s		m	
vibrational coordinates,				9
internal	R_i, r_i, θ_j, etc.		(varies)	
symmetry	S_i		(varies)	
normal				
mass adjusted	Q_r		$\mathrm{kg}^{\frac{1}{2}}$ m	
dimensionless	q_r		1	

(6) d is usually used for vibrational degeneracy, and β for nuclear spin degeneracy.
(7) Molecular dipole moments are often expressed in the non-SI unit debye, where $\mathrm{D} \approx 3.335\,64 \times 10^{-30}$ C m. The SI unit C m is inconvenient for expressing molecular dipole moments, which results in the continued use of the deprecated debye (D). A convenient alternative is to use the atomic unit, ea_0. Another way of expressing dipole moments is to quote the electric dipole lengths, $l_p = p/e$, analogous to the way the nuclear quadrupole areas are quoted (see pp.21 and 98). This gives the distance between two elementary charges of the equivalent dipole and conveys a clear picture in relation to molecular dimensions.

Examples

	Dipole moment			Dipole length
	SI		a.u.	
	p/C m	p/D	p/ea_0	l_p/pm
HCl	3.60×10^{-30}	1.08	0.425	22.5
H_2O	6.23×10^{-30}	1.87	0.736	38.9
NaCl	4.02×10^{-29}	12.1	4.74	251

See also footnote (4) on p.14.
(8) For quantities describing line and band intensities see section 2.7, p.33–35.
(9) Interatomic (internuclear) distances and vibrational displacements are often expressed in the non-SI unit ångström, where $\mathrm{Å} = 10^{-10}\ \mathrm{m} = 0.1\ \mathrm{nm} = 100\ \mathrm{pm}$.
(10) The various slightly different ways of representing interatomic distances, distinguished by subscripts, involve different vibrational averaging contributions; they are discussed in [54], where the geometrical structures of many free molecules are listed. Only the equilibrium distance r_e is isotopically invariant. The effective distance parameter r_0 is estimated from the rotational constants for the ground vibrational state and has only approximate physical significance for polyatomic molecules.

Name	*Symbol*	*Definition*	*SI unit*	*Notes*
vibrational force constants,				
diatomic	$f, (k)$	$f = \partial^2 V/\partial r^2$	$\mathrm{J\ m^{-2}}$	11
polyatomic,				
internal coordinates	f_{ij}	$f_{ij} = \partial^2 V/\partial r_i \partial r_j$	(varies)	
symmetry coordinates	F_{ij}	$F_{ij} = \partial^2 V/\partial S_i \partial S_j$	(varies)	
dimensionless normal coordinates	$\phi_{rst\ldots}, k_{rst\ldots}$		$\mathrm{m^{-1}}$	12
nuclear magnetic resonance (NMR):				
magnetogyric ratio	γ	$\gamma = \mu/I\hbar$	$\mathrm{s^{-1}\ T^{-1}}$	
shielding constant	σ	$B_A = (1 - \sigma_A)B$	1	13
chemical shift, δ scale	δ	$\delta = 10^6(\nu - \nu_0)/\nu_0$	1	14
coupling constant,				
(indirect) spin–spin	J_{AB}	$\hat{H}/h = J_{AB}\hat{\boldsymbol{I}}_A \cdot \hat{\boldsymbol{I}}_B$	Hz	15
reduced spin–spin	K_{AB}	$K_{AB} = \dfrac{J_{AB}}{h}\dfrac{2\pi}{\gamma_A}\dfrac{2\pi}{\gamma_B}$	$\mathrm{T^2\ J^{-1}}$, $\mathrm{N\ A^{-2}\ m^{-3}}$	16
direct (dipolar)	D_{AB}		Hz	17
relaxation time,				
longitudinal	T_1		s	18
transverse	T_2		s	18

(11) Force constants are often expressed in mdyn Å^{-1} = aJ Å^{-2} for stretching coordinates, mdyn Å = aJ for bending coordinates, and mdyn = aJ Å^{-1} for stretch–bend interactions. See [17] for further details on definitions and notation for force constants.

(12) The force constants in dimensionless normal coordinates are usually defined in wavenumber units by the equation $V/hc = \Sigma \phi_{rst\ldots}\, q_r q_s q_t \ldots$, where the summation over the normal coordinate indices $r, s, t, \ldots$ is unrestricted.

(13) σ_A and B_A denote the shielding constant and the local magnetic field at nucleus A.

(14) ν_0 is the resonance frequency of a reference molecule, usually tetramethylsilane for proton and for ^{13}C resonance spectra [12]. In some of the older literature proton chemical shifts are expressed on the τ scale, where $\tau = 10 - \delta$, but this is no longer used.

(15) $\hat{H}$ in the definition is the spin–spin coupling hamiltonian between nuclei A and B.

(16) Whereas J_{AB} involves the nuclear magnetogyric ratios, the reduced coupling constant K_{AB} represents only the electronic contribution and is thus approximately isotope independent and may exhibit chemical trends.

(17) Direct dipolar coupling occurs in solids; the definition of the coupling constant is $D_{AB} = (\mu_0/4\pi) r_{AB}{}^{-3} \gamma_A \gamma_B (\hbar/2\pi)$.

(18) The longitudinal relaxation time is associated with spin–lattice relaxation, and the transverse relaxation time with spin–spin relaxation. The definitions are

$$\mathrm{d}M_z/\mathrm{d}t = -(M_z - M_{z,e})/T_1,$$

and

$$\mathrm{d}M_x/\mathrm{d}t = -M_x/T_2,$$

where M_z and M_x are the components of magnetization parallel and perpendicular to the static field B, and $M_{z,e}$ is the equilibrium value of M_z.

Name	Symbol	Definition	SI unit	Notes
electron spin resonance (ESR),				
electron paramagnetic resonance (EPR):				
magnetogyric ratio	γ	$\gamma = \mu/s\hbar$	$s^{-1} T^{-1}$	
g-factor	g	$h\nu = g\mu_B B$	1	
hyperfine coupling constant,				
in liquids	a, A	$\hat{H}_{hfs}/h = a\hat{\boldsymbol{S}}\cdot\hat{\boldsymbol{I}}$	Hz	19
in solids	$\boldsymbol{T}$	$\hat{H}_{hfs}/h = \hat{\boldsymbol{S}}\cdot\boldsymbol{T}\cdot\hat{\boldsymbol{I}}$	Hz	19

(19) $\hat{H}_{hfs}$ is the hyperfine coupling hamiltonian. The coupling constants a are usually quoted in MHz, but they are sometimes quoted in magnetic induction units (G or T) obtained by dividing by the conversion factor $g\mu_B/h$, which has the SI unit Hz/T; $g_e\mu_B/h \approx 28.025$ GHz T^{-1} (= 2.8025 MHz G^{-1}), where g_e is the g-factor for a free electron. If in liquids the hyperfine coupling is isotropic, the coupling constant is a scalar a. In solids the coupling is anisotropic, and the coupling constant is a 3×3 tensor $\boldsymbol{T}$. Similar comments apply to the g-factor.

Symbols for angular momentum operators and quantum numbers

In the following table, all of the operator symbols denote the dimensionless ratio (*angular momentum*)/$\hbar$. (Although this is a universal practice for the quantum numbers, some authors use the operator symbols to denote *angular momentum*, in which case the operators would have SI units: J s.) The column heading 'Z-axis' denotes the space-fixed component, and the heading 'z-axis' denotes the molecule-fixed component along the symmetry axis (linear or symmetric top molecules), or the axis of quantization.

Angular momentum[1]	*Operator symbol*	*Quantum number symbol* *Total*	*Z-axis*	*z-axis*	*Notes*
electron orbital	$\hat{\boldsymbol{L}}$	L	M_L	Λ	2
one electron only	$\hat{\boldsymbol{l}}$	l	m_l	λ	2
electron spin	$\hat{\boldsymbol{S}}$	S	M_S	Σ	
one electron only	$\hat{\boldsymbol{s}}$	s	m_s	σ	
electron orbital + spin	$\hat{\boldsymbol{L}} + \hat{\boldsymbol{S}}$			$\Omega = \Lambda + \Sigma$	2
nuclear orbital (rotational)	$\hat{\boldsymbol{R}}$	R		K_R, k_R	
nuclear spin	$\hat{\boldsymbol{I}}$	I	M_I		
internal vibrational					
spherical top	$\hat{\boldsymbol{l}}$	$l(l\zeta)$		K_l	3
other	$\hat{\boldsymbol{j}}, \hat{\boldsymbol{\pi}}$			$l(l\zeta)$	2, 3
sum of $R + L(+j)$	$\hat{\boldsymbol{N}}$	N		K, k	2
sum of $N + S$	$\hat{\boldsymbol{J}}$	J	M_J	K, k	2, 4
sum of $J + I$	$\hat{\boldsymbol{F}}$	F	M_F		

(1) In all cases the vector operator and its components are related to the quantum numbers by eigenvalue equations analogous to:

$$\hat{J}^2\psi = J(J+1)\psi,\ \hat{J}_Z\psi = M_J\psi, \text{ and } \hat{J}_z\psi = K\psi,$$

where the component quantum numbers M_J and K take integral or half-odd values in the range $-J \leqslant M_J \leqslant +J$, $-J \leqslant K \leqslant +J$. (If the operator symbols are taken to represent *angular momentum*, rather

Symbols for symmetry operators and labels for symmetry species

(i) Symmetry operators in space-fixed coordinates [55]

identity	E
permutation	P
space-fixed inversion	E^*
permutation-inversion	P^* $(= PE^*)$

The permutation operation P permutes the labels of identical nuclei.

Example In the NH_3 molecule, if the hydrogen nuclei are labelled 1, 2 and 3, then $P = (123)$ would symbolize the permutation 1 is replaced by 2, 2 by 3, and 3 by 1.

The inversion operation E^* reverses the sign of all particle coordinates in the space-fixed origin, or in the molecule-fixed centre of mass if translation has been separated. It is also called the parity operator; in field-free space, wavefunctions are either parity + (unchanged) or parity − (change sign) under E^*. The label may be used to distinguish the two nearly degenerate components formed by Λ-doubling (in a degenerate electronic state) or l-doubling (in a degenerate vibrational state) in linear molecules, or by K-doubling (asymmetry-doubling) in slightly asymmetric tops. For linear molecules, Λ- or l-doubled components may also be distinguished by the labels e or f [56]; for singlet states these correspond respectively to parity + or − for J even and vice versa for J odd (but see [56]). For linear molecules in degenerate electronic states the Λ-doubled levels may alternatively be labelled $\Pi(A')$ or $\Pi(A'')$ (or $\Delta(A')$, $\Delta(A'')$, etc.) [57]. Here the labels A′ or A″ describe the symmetry of the electronic wavefunction at high J with respect to reflection in the plane of rotation (but see [57] for further details). The A′ or A″ labels are particularly useful for the correlation of states of molecules involved in reactions or photodissociation.

In relation to permutation inversion symmetry species the superscript + or − may be used to designate parity.

Examples: A_1^+ totally symmetric with respect to permutation, positive parity
A_1^- totally symmetric with respect to permutation, negative parity

The Herman–Maugin symbols of symmetry operations used for crystals are given in section 2.8 on p.38.

Notes (continued)

than (*angular momentum*)/$\hbar$, the eigenvalue equations should read $\hat{J}^2\psi = J(J+1)\hbar^2\psi$, $\hat{J}_Z\psi = M_J\hbar\psi$, and $\hat{J}_z\psi = K\hbar\psi$.)

(2) Some authors, notably Herzberg [52], treat the component quantum numbers Λ, Ω, l and K as taking positive or zero values only, so that each non-zero value of the quantum number labels two wavefunctions with opposite signs for the appropriate angular momentum component. When this is done, lower case k is often regarded as a signed quantum number, related to K by $K = |k|$. However, in theoretical discussions all component quantum numbers are usually treated as signed, taking both positive and negative values.

(3) There is no uniform convention for denoting the internal vibrational angular momentum; j, π, p and G have all been used. For symmetric top and linear molecules the component of j in the symmetry axis is always denoted by the quantum number l, where l takes values in the range $-v \leqslant l \leqslant +v$ in steps of 2. The corresponding component of angular momentum is actually $l\zeta\hbar$, rather than $l\hbar$, where ζ is a Coriolis coupling constant.

(4) Asymmetric top rotational states are labelled by the value of J (or N if $S \neq 0$), with subscripts K_a, K_c, where the latter correlate with the $K = |k|$ quantum number about the a and c axes in the prolate and oblate symmetric top limits respectively.

Example $J_{K_a,K_c} = 5_{2,3}$ for a particular rotational level.

(ii) Symmetry operators in molecule-fixed coordinates (Schönflies symbols) [52]

identity	E
rotation by $2\pi/n$	C_n
reflection	$\sigma, \sigma_v, \sigma_d, \sigma_h$
inversion	i
rotation-reflection	$S_n\ (= C_n\sigma_h)$

If C_n is the primary axis of symmetry, wavefunctions that are unchanged or change sign under the operator C_n are given species labels A or B respectively, and otherwise wavefunctions that are multiplied by $\exp(\pm 2\pi i s/n)$ are given the species label E_s. Wavefunctions that are unchanged or change sign under i are labelled g (gerade) or u (ungerade) respectively. Wavefunctions that are unchanged or change sign under σ_h have species labels with a ′ or ″ respectively. For more detailed rules see [51, 52].

Other symbols and conventions in optical spectroscopy

(i) Term symbols for atomic states

The electronic states of atoms are labelled by the value of the quantum number L for the state. The value of L is indicated by an upright capital letter: S, P, D, F, G, H, I, and K, . . . , are used for $L = 0$, 1, 2, 3, 4, 5, 6, and 7, . . . , respectively. The corresponding lower case letters are used for the orbital angular momentum of a single electron. For a many-electron atom, the electron spin multiplicity $(2S + 1)$ may be indicated as a left-hand superscript to the letter, and the value of the total angular momentum J as a right-hand subscript. If either L or S is zero only one value of J is possible, and the subscript is then usually suppressed. Finally, the electron configuration of an atom is indicated by giving the occupation of each one-electron orbital as in the examples below.

Examples B: $(1s)^2(2s)^2(2p)^1$, ${}^2P_{1/2}$
C: $(1s)^2(2s)^2(2p)^2$, ${}^3P_0$
N: $(1s)^2(2s)^2(2p)^3$, ${}^4S$

(ii) Term symbols for molecular states

The electronic states of molecules are labelled by the symmetry species label of the wavefunction in the molecular point group. These should be Latin or Greek upright capital letters. As for atoms, the spin multiplicity $(2S + 1)$ may be indicated by a left superscript. For linear molecules the value of $\Omega\ (= \Lambda + \Sigma)$ may be added as a right subscript (analogous to J for atoms). If the value of Ω is not specified, the term symbol is taken to refer to all component states, and a right subscript r or i may be added to indicate that the components are regular (energy increases with Ω) or inverted (energy decreases with Ω) respectively.

The electronic states of molecules are also given empirical single letter labels as follows. The ground electronic state is labelled X, excited states of the same multiplicity are labelled A, B, C, . . . , in ascending order of energy, and excited states of different multiplicity are labelled with lower case letters a, b, c, ... In polyatomic molecules (but not diatomic molecules) it is customary to add a tilde (e.g. $\tilde{X}$) to these empirical labels to prevent possible confusion with the symmetry species label.

Finally the one-electron orbitals are labelled by the corresponding lower case letters, and the electron configuration is indicated in a manner analogous to that for atoms.

Examples The ground state of CH is $(1\sigma)^2(2\sigma)^2(3\sigma)^2(1\pi)^1$, X ${}^2\Pi_r$, in which the ${}^2\Pi_{1/2}$ component lies below the ${}^2\Pi_{3/2}$ component, as indicated by the subscript r for regular.

The ground state of OH is $(1\sigma)^2(2\sigma)^2(3\sigma)^2(1\pi)^3$, X ${}^2\Pi_i$, in which the ${}^2\Pi_{3/2}$ component lies below the ${}^2\Pi_{1/2}$ component, as indicated by the subscript i for inverted.

The two lowest electronic states of CH_2 are . . . $(2a_1)^2(1b_2)^2(3a_1)^2$, $\tilde{a}\ ^1A_1$,
. . . $(2a_1)^2(1b_2)^2(3a_1)^1(1b_1)^1$, $\tilde{X}\ ^3B_1$.

The ground state of C_6H_6 (benzene) is . . . $(a_{2u})^2(e_{1g})^4$, $\tilde{X}\ ^1A_{1g}$.

The vibrational states of molecules are usually indicated by giving the vibrational quantum numbers for each normal mode.

Examples For a bent triatomic molecule,
(0, 0, 0) denotes the ground state,
(1, 0, 0) denotes the ν_1 state, i.e. $v_1 = 1$, and
(1, 2, 0) denotes the $\nu_1 + 2\nu_2$ state, etc.

(iii) Notation for spectroscopic transitions

The upper and lower levels of a spectroscopic transition are indicated by a prime $'$ and double-prime $''$ respectively.

Example $h\nu = E' - E''$

Transitions are generally indicated by giving the excited state label, followed by the ground state label, separated by a dash or an arrow to indicate the direction of the transition (emission to the right, absorption to the left).

Examples B – A indicates a transition between a higher energy state B and a lower energy state A;
B → A indicates emission from B to A;
B ← A indicates absorption from A to B.
(0, 2, 1) ← (0, 0, 1) labels the $2\nu_2 + \nu_3 - \nu_3$ hot band in a bent triatomic molecule.

A more compact notation [58] may be used to label vibronic (or vibrational) transitions in polyatomic molecules with many normal modes, in which each vibration index r is given a superscript v_r' and a subscript v_r'' indicating the upper and lower state values of the quantum number. When $v_r' = v_r'' = 0$ the corresponding index is suppressed.

Examples 1_0^1 denotes the transition (1, 0, 0) – (0, 0, 0);
$2_0^2\ 3_1^1$ denotes the transition (0, 2, 1) – (0, 0, 1).

For rotational transitions, the value of $\Delta J = J' - J''$ is indicated by a letter labelling the branches of a rotational band: $\Delta J = -2, -1, 0, 1,$ and 2 are labelled as the O-branch, P-branch, Q-branch, R-branch, and S-branch respectively. The changes in other quantum numbers (such as K for a symmetric top, or K_a and K_c for an asymmetric top) may be indicated by adding lower case letters as a left superscript according to the same rule.

Example ^{p}Q labels a 'p-type Q-branch' in a symmetric top molecule, i.e. $\Delta K = -1$, $\Delta J = 0$.

(iv) Presentation of spectra

It is recommended to plot both infrared and visible/ultraviolet spectra against wavenumber, usually in cm^{-1}, with decreasing wavenumber to the right (note the mnemonic 'red to the right', derived for the visible region) [10, 18]. (Visible/ultraviolet spectra are also sometimes plotted against wavelength, usually in nm, with increasing wavelength to the right.) It is recommended to plot Raman spectra with increasing wavenumber shift to the left [11].

It is recommended to plot both electron spin resonance (ESR) spectra and nuclear magnetic resonance (NMR) spectra with increasing magnetic induction (loosely called magnetic field) to the right for fixed frequency, or with increasing frequency to the left for fixed magnetic field [12, 13].

It is recommended to plot photoelectron spectra with increasing ionization energy to the left, i.e. with increasing photoelectron kinetic energy to the right [14].

2.7 ELECTROMAGNETIC RADIATION

The quantities and symbols given here have been selected on the basis of recommendations by IUPAP [4], ISO [5.g], and IUPAC [19–21] as well as by taking into account the practice in the field of laser physics.

Name	*Symbol*	*Definition*	*SI unit*	*Notes*
wavelength	λ		m	
speed of light				
in vacuum	c_0	$c_0 = 299\,792\,458\ \mathrm{m\,s^{-1}}$	$\mathrm{m\,s^{-1}}$	1
in a medium	c	$c = c_0/n$	$\mathrm{m\,s^{-1}}$	
wavenumber in vacuum	$\tilde{\nu}$	$\tilde{\nu} = \nu/c_0 = 1/n\lambda$	$\mathrm{m^{-1}}$	2
wavenumber (in a medium)	σ	$\sigma = 1/\lambda$	$\mathrm{m^{-1}}$	
frequency	ν	$\nu = c/\lambda$	Hz	
angular frequency, pulsatance	ω	$\omega = 2\pi\nu$	$\mathrm{s^{-1}}$, $\mathrm{rad\,s^{-1}}$	
refractive index	n	$n = c_0/c$	1	
Planck constant	h		J s	
Planck constant/2π	$\hbar$	$\hbar = h/2\pi$	J s	
radiant energy	Q, W		J	3
radiant energy density	ρ, w	$\rho = Q/V$	$\mathrm{J\,m^{-3}}$	3
spectral radiant energy density				3
in terms of frequency	ρ_ν, w_ν	$\rho_\nu = \mathrm{d}\rho/\mathrm{d}\nu$	$\mathrm{J\,m^{-3}\,Hz^{-1}}$	
in terms of wavenumber	$\rho_{\tilde{\nu}}$, $w_{\tilde{\nu}}$	$\rho_{\tilde{\nu}} = \mathrm{d}\rho/\mathrm{d}\tilde{\nu}$	$\mathrm{J\,m^{-2}}$	
in terms of wavelength	ρ_λ, w_λ	$\rho_\lambda = \mathrm{d}\rho/\mathrm{d}\lambda$	$\mathrm{J\,m^{-4}}$	
Einstein transition probabilities,				4, 5
spontaneous emission	A_{ij}	$\mathrm{d}N_j/\mathrm{d}t = -\sum_i A_{ij}N_j$	$\mathrm{s^{-1}}$	

(1) When there is no risk of ambiguity the subscript $_0$ denoting vacuum is often omitted.
(2) The unit $\mathrm{cm^{-1}}$ is generally used for wavenumber in vacuum.
(3) The symbols for the quantities *radiant energy* through *irradiance* are also used for the corresponding quantities concerning visible radiation, i.e. luminous quantities and photon quantities. Subscripts e for energetic, v for visible, and p for photon may be added whenever confusion between these quantities might otherwise occur. The units used for luminous quantities are derived from the base unit candela (cd), see chapter 3.

Example radiant intensity I_e, SI unit: $\mathrm{W\,sr^{-1}}$
luminous intensity I_v, SI unit: cd
photon intensity I_p, SI units: $\mathrm{s^{-1}\,sr^{-1}}$

(4) The indices i and j refer to individual states; $E_j > E_i$, $E_j - E_i = hc\tilde{\nu}_{ij}$, and $B_{ji} = B_{ij}$ in the defining equations. The coefficients B are defined here using energy density $\rho_{\tilde{\nu}}$ in terms of wavenumber; they may alternatively be defined using energy density in terms of frequency ρ_ν, in which case B has SI units $\mathrm{m\,kg^{-1}}$, and $B_\nu = c_0 B_{\tilde{\nu}}$ where B_ν is defined using frequency and $B_{\tilde{\nu}}$ using wavenumber.
(5) The relation between the Einstein coefficients A and $B_{\tilde{\nu}}$ is $A = 8\pi hc_0\tilde{\nu}^3 B_{\tilde{\nu}}$. The Einstein stimulated absorption or emission coefficient B may also be related to the transition moment between the states i and j; for an electric dipole transition the relation is

$$B_{\tilde{\nu},ij} = \frac{8\pi^3}{3h^2 c_0 (4\pi\varepsilon_0)} \sum_\rho |\langle i|\mu_\rho|j\rangle|^2$$

where the sum over ρ goes over the three space-fixed cartesian axes, and μ_ρ is a space-fixed component of the dipole moment operator. Again, these equations are based on a wavenumber definition of the Einstein coefficient B (i.e. $B_{\tilde{\nu}}$ rather than B_ν).

Name	*Symbol*	*Definition*	*SI unit*	*Notes*
Einstein transition probabilities (cont.)				4, 5
stimulated emission, induced emission	B_{ij}	$\mathrm{d}N_j/\mathrm{d}t = -\sum_i \rho_{\tilde{\nu}}(\tilde{\nu}_{ij})B_{ij}N_j$	$\mathrm{s\ kg^{-1}}$	
absorption	B_{ji}	$\mathrm{d}N_i/\mathrm{d}t = -\sum_j \rho_{\tilde{\nu}}(\tilde{\nu}_{ij})B_{ji}N_i$	$\mathrm{s\ kg^{-1}}$	
radiant power, radiant energy per time	Φ, P	$\Phi = \mathrm{d}Q/\mathrm{d}t$	W	3
radiant intensity	I	$I = \mathrm{d}\Phi/\mathrm{d}\Omega$	$\mathrm{W\ sr^{-1}}$	3
radiant excitance (emitted radiant flux)	M	$M = \mathrm{d}\Phi/\mathrm{d}A_{\mathrm{source}}$	$\mathrm{W\ m^{-2}}$	3
radiance	L	$L = \dfrac{\mathrm{d}^2\Phi}{\mathrm{d}\Omega\,\mathrm{d}A_{\mathrm{source}}}$	$\mathrm{W\ sr^{-1}\ m^{-2}}$	3, 6
intensity, irradiance (radiant flux received)	I, E	$I = \mathrm{d}\Phi/\mathrm{d}A$	$\mathrm{W\ m^{-2}}$	3, 7
spectral intensity, spectral irradiance	$I(\tilde{\nu})$, $E(\tilde{\nu})$	$I(\tilde{\nu}) = \mathrm{d}I/\mathrm{d}\tilde{\nu}$	$\mathrm{W\ m^{-1}}$	8
fluence	F, (H)	$F = \int I\,\mathrm{d}t = \int \dfrac{\mathrm{d}\Phi}{\mathrm{d}A}\mathrm{d}t$	$\mathrm{J\ m^{-2}}$	9
emittance	ε	$\varepsilon = M/M_{\mathrm{bb}}$	1	10
Stefan–Boltzmann constant	σ	$M_{\mathrm{bb}} = \sigma T^4$	$\mathrm{W\ m^{-2}\ K^{-4}}$	10
étendue (throughput, light gathering power)	E, (e)	$E = A\Omega = \Phi/L$	$\mathrm{m^2\ sr}$	11
resolving power	R	$R = \tilde{\nu}/\delta\tilde{\nu}$	1	12
resolution	$\delta\tilde{\nu}$		$\mathrm{m^{-1}}$	2, 12, 13
free spectral range	$\Delta\tilde{\nu}$	$\Delta\tilde{\nu} = 1/2l$	$\mathrm{m^{-1}}$	2, 14
finesse	f	$f = \Delta\tilde{\nu}/\delta\tilde{\nu}$	1	14
quality factor	Q	$Q = 2\pi\nu \dfrac{W}{-\mathrm{d}W/\mathrm{d}t}$	1	14, 15

(6) The radiance is a normalized measure of the brightness of a source; it is the power emitted per area of source. per solid angle of the beam from each point of the source.
(7) The name intensity, symbol I, is usually used in discussions involving collimated beams of light, as in applications of the Lambert–Beer law for spectrometric analysis.
(8) Spectral quantities may also be defined with respect to frequency ν, or wavelength λ; see spectral radiant energy density above.
(9) Fluence is used in photochemistry to specify the energy delivered in a given time interval (for instance by a laser pulse). This quantity may also be called radiant exposure.
(10) The emittance of a sample is the ratio of the flux emitted by the sample to the flux emitted by a black body at the same temperature; M_{bb} is the latter quantity.
(11) Étendue is a characteristic of an optical instrument. It is a measure of the light gathering power, i.e. the power transmitted per radiance of the source. A is the area of the source (or image stop); Ω is the solid angle accepted from each point of the source by the aperture stop.
(12) This quantity characterizes the performance of a spectrometer, or the degree to which a spectral line (or a laser beam) is monochromatic. It may also be defined using frequency ν, or wavelength λ.
(13) The precise definition of resolution depends on the lineshape, but usually resolution is taken as the full line width at half maximum intensity (FWHM) on a wavenumber, $\delta\tilde{\nu}$, or frequency, $\delta\nu$, scale.
(14) These quantities characterize a Fabry–Perot cavity, or a laser cavity. l is the cavity spacing, and $2l$ is the round-trip path length. The free spectral range is the wavenumber interval between successive longitudinal cavity modes.
(15) W is the energy stored in the cavity, and $-\mathrm{d}W/\mathrm{d}t$ is the rate of decay of stored energy. Q is also related to the linewidth of a single cavity mode: $Q = \nu/\delta\nu = \tilde{\nu}/\delta\tilde{\nu}$. Thus high Q cavities give narrow linewidths.

Name	*Symbol*	*Definition*	*SI unit*	*Notes*
first radiation constant	c_1	$c_1 = 2\pi hc_0^2$	$\mathrm{W\,m^2}$	
second radiation constant	c_2	$c_2 = hc_0/k$	$\mathrm{K\,m}$	
transmittance, transmission factor	τ, T	$\tau = \Phi_{tr}/\Phi_0$	1	16, 17
absorptance, absorption factor	α	$\alpha = \Phi_{abs}/\Phi_0$	1	16, 17
reflectance, reflection factor	ρ	$\rho = \Phi_{refl}/\Phi_0$	1	16, 17
(decadic) absorbance	A_{10}, A	$A_{10} = -\lg(1-\alpha_i)$	1	17, 18, 19
napierian absorbance	A_e, B	$A_e = -\ln(1-\alpha_i)$	1	17, 18, 19
absorption coefficient,				
(linear) decadic	a, K	$a = A_{10}/l$	$\mathrm{m^{-1}}$	17, 20
(linear) napierian	α	$\alpha = A_e/l$	$\mathrm{m^{-1}}$	17, 20
molar (decadic)	ε	$\varepsilon = a/c = A_{10}/cl$	$\mathrm{m^2\,mol^{-1}}$	17, 20, 21
molar napierian	κ	$\kappa = \alpha/c = A_e/cl$	$\mathrm{m^2\,mol^{-1}}$	17, 20, 21
net absorption cross section	σ_{net}	$\sigma_{net} = \kappa/N_A$	$\mathrm{m^2}$	22
integrated absorption intensity				
—against $\tilde{\nu}$	$A, \bar{A}$	$A = \int \kappa(\tilde{\nu})\,\mathrm{d}\tilde{\nu}$	$\mathrm{m\,mol^{-1}}$	22, 23
	S	$S = A/N_A$	m	22, 23
	$\bar{S}$	$\bar{S} = (1/pl)\int \ln(I_0/I)\,\mathrm{d}\tilde{\nu}$	$\mathrm{Pa^{-1}\,m^{-2}}$	22, 23, 24

(16) If scattering and luminescence can be neglected, $\tau + \alpha + \rho = 1$. In optical spectroscopy internal properties (denoted by subscript i) are defined to exclude surface effects and effects of the cuvette such as reflection losses, so that if scattering and luminescence in the sample can be neglected $\tau_i + \alpha_i = 1$. This leads to the customary form of the Lambert–Beer law, $\Phi_{tr}/\Phi_0 = I_{tr}/I_0 = \tau_i = 1 - \alpha_i = \exp(-\kappa cl)$.
(17) In spectroscopy all of these quantities are usually taken to be defined in terms of the spectral intensity, $I(\tilde{\nu})$, so that they are all regarded as functions of wavenumber $\tilde{\nu}$ (or frequency ν) across the spectrum. Thus, for example, the absorption coefficient $\alpha(\tilde{\nu})$ at wavenumber $\tilde{\nu}$ defines the absorption spectrum of the sample; similarly $T(\tilde{\nu})$ defines the transmittance spectrum.
(18) The definitions given here relate the absorbance A_{10} or A_e to the *internal* absorptance α_i; see note (16). However the subscript i on the absorptance α is often omitted.
(19) In reference [19] the symbol A is used for decadic absorbance, and B for napierian absorbance.
(20) l is the absorbing path length, and c is the amount (of substance) concentration.
(21) The molar decadic absorption coefficient ε is frequently called the 'extinction coefficient' in published literature. Unfortunately numerical values of the 'extinction coefficient' are often quoted without specifying units; the absence of units usually means that the units are $\mathrm{mol^{-1}\,dm^3\,cm^{-1}}$. See also [18]. The word 'extinction' should properly be reserved for the sum of the effects of absorption, scattering, and luminescence.
(22) Note that these quantities give the net absorption coefficient κ, the net absorption cross section σ_{net}, and the net values of A, S, $\bar{S}$, Γ, and G_{net}, in the sense that they are the sums of effects due to absorption and induced emission. See the discussion below on p.33–34.
(23) The definite integral defining these quantities may be specified by the limits of integration in parentheses, e.g. $G(\tilde{\nu}_1, \tilde{\nu}_2)$. In general the integration is understood to be taken over an absorption line or an absorption band. A, $\bar{S}$, and Γ are measures of the strength of the band in terms of amount concentration; $G_{net} = \Gamma/N_A$ and $S = A/N_A$ are corresponding molecular quantities. For a single spectral line the relation of these quantities to the Einstein transition probabilities is discussed below on p.34. The symbol $\bar{A}$ may be used for the integrated absorption coefficient A when there is a possibility of confusion with the Einstein spontaneous emission coefficient A_{ij}.

The integrated intensity of an electronic transition is often expressed in terms of the oscillator strength or 'f value', which is dimensionless, or in terms of the Einstein transition probability A_{ij} between the states involved,

Name	*Symbol*	*Definition*	*SI unit*	*Notes*
Integrated absorption intensities (cont.)				
—against $\ln \tilde{\nu}$	Γ	$\Gamma = \int \kappa(\tilde{\nu})\, \tilde{\nu}^{-1}\, d\tilde{\nu}$	$m^2\, mol^{-1}$	22, 23
integrated absorption cross section	G_{net}	$G_{net} = \int \sigma_{net}(\tilde{\nu})\, \tilde{\nu}^{-1}\, d\tilde{\nu}$	m^2	22, 23
absorption index	k	$k = \alpha/4\pi\tilde{\nu}$	1	25
complex refractive index	$\hat{n}$	$\hat{n} = n + ik$	1	
molar refraction	R	$R = \left(\frac{n^2 - 1}{n^2 + 2}\right) V_m$	$m^3\, mol^{-1}$	
angle of optical rotation	α		1, rad	26
specific optical rotatory power	$[\alpha]^{\theta}_{\lambda}$	$[\alpha]^{\theta}_{\lambda} = \alpha/\gamma l$	$rad\, m^2\, kg^{-1}$	26
molar optical rotatory power	α_m	$\alpha_m = \alpha/cl$	$rad\, m^2\, mol^{-1}$	26

with SI unit s^{-1}. Whereas A_{ij} has a simple and universally accepted meaning (see p.30), there are differing uses of f. A common practical conversion is given by the equation

$$f_{ij} = [(4\pi\varepsilon_0)m_e c_0/8\pi^2 e^2]\lambda^2 A_{ij}, \quad \text{or} \quad f_{ij} = (1.4992 \times 10^{-14})(A_{ij}/s^{-1})(\lambda/nm)^2,$$

where λ is the transition wavelength, and i and j refer to individual states. For strongly allowed electronic transitions f is of the order unity.

(24) The quantity $\bar{S}$ is only used for gases; it is defined in a manner similar to A, except that the partial pressure of gas p replaces the concentration c. At low pressures $p_i \approx c_i RT$, so that $\bar{S}$ and A are related by the equation $\bar{S} \approx A/RT$. Thus if $\bar{S}$ is used to report line or band intensities, the temperature should be specified.

(25) α in the definition is the napierian absorption coefficient.

(26) The sign convention for the angle of rotation is as follows: α is positive if the plane of polarization is rotated clockwise as viewed looking towards the light source. If the rotation is anticlockwise then α is negative.

The optical rotation due to a solute in solution may be specified by a statement of the type

$$\alpha(589.3\text{ nm}, 20\,^{\circ}\text{C}, \text{sucrose}, 10\,\text{g dm}^{-3} \text{ in } H_2O, 10\text{ cm path}) = +0.6647^{\circ}$$

The same information may be conveyed by quoting either the specific optical rotatory power $\alpha/\gamma l$, or the molar optical rotatory power α/cl, where γ is the mass concentration, c is the amount (of substance) concentration, and l is the path length. Most tabulations give the specific optical rotatory power, denoted $[\alpha]^{\theta}_{\lambda}$. The wavelength of light used λ (frequently the sodium D line) and the Celsius temperature θ are conventionally written as a subscript and superscript to the specific rotatory power $[\alpha]$. For pure liquids and solids $[\alpha]^{\theta}_{\lambda}$ is similarly defined as $[\alpha]^{\theta}_{\lambda} = \alpha/\rho l$, where ρ is the mass density.

Specific optical rotatory powers are customarily called *specific rotations*, and are unfortunately usually quoted without units. The absence of units may usually be taken to mean that the units are $deg\, cm^3\, g^{-1}\, dm^{-1}$ for pure liquids and solutions, or $deg\, cm^3\, g^{-1}\, mm^{-1}$ for solids, where deg is used as a symbol for degrees of plane angle.

Quantities and symbols concerned with the measurement of absorption intensity

In most experiments designed to measure the intensity of spectral absorption, the measurement gives the net absorption due to the effects of absorption from the lower energy level m to the upper energy level n, less induced emission from n to m. Since the populations depend on the temperature, so will the measured net absorption. This comment applies to all the quantities defined in the table to measure absorption intensity, although for transitions where $hc_0\tilde{\nu} \gg kT$ the temperature dependence is small and for $\tilde{\nu} > 1000\ cm^{-1}$ induced emission can generally be neglected.

In a more fundamental approach one defines the pure absorption cross section $\sigma_{ji}(\tilde{\nu})$ for an induced radiative transition from the state i to the state j (in either absorption or emission). For an ideal absorption experiment with only the lower state i populated the integrated absorption cross section for the transition $j \leftarrow i$ is given by

$$G_{ji} = \int \sigma_{ji}(\tilde{\nu})\tilde{\nu}^{-1}\, d\tilde{\nu} = \int \sigma_{ji}(\nu)\,\nu^{-1}\, d\nu$$

If the upper and lower energy levels are degenerate the observed line strength is given by summing over transitions between all states i in the lower energy level m and all states j in the upper energy level n, multiplying each term by the fractional population p_i in the appropriate initial state. Neglecting induced emission this gives

$$G_{\text{net}}(n \leftarrow m) = \sum_{i,j} p_i G_{ji}$$

If induced emission is significant then the net integrated cross section will be

$$G_{\text{net}}(n \leftarrow m) = \sum_{i,\,j} (p_i - p_j)G_{ji} = (p_m/d_m - p_n/d_n)\sum_{i,\,j} G_{ji}$$

Here p_i and p_j denote the fractional populations of states i and j ($p_i = \exp\{-E_i/kT\}/q$ in thermal equilibrium, where q is the partition function); p_m and p_n denote the corresponding fractional populations of the energy levels, and d_m and d_n the degeneracies ($p_i = p_m/d_m$, etc.). The absorption intensity G_{ji}, and the Einstein coefficients A_{ij} and B_{ji}, are fundamental measures of the line strength between the individual states i and j; they are related to each other by the general equations

$$G_{ji} = hB_{\tilde{\nu},\,ji} = (h/c_0)B_{\nu,\,ji} = A_{ij}/8\pi c_0\tilde{\nu}^3$$

Finally, for an electric dipole transition these quantities are related to the square of the transition moment by the equation

$$G_{ji} = hB_{\tilde{\nu},\,ji} = A_{ij}/8\pi c_0\tilde{\nu}^3 = \frac{8\pi^3}{3hc_0(4\pi\varepsilon_0)}|\boldsymbol{M}_{ji}|^2$$

where the transition moment $\boldsymbol{M}_{ji}$ is given by

$$|\boldsymbol{M}_{ji}|^2 = \sum_{\rho} |\langle i|\mu_\rho|j\rangle|^2$$

Here the sum is over the three space-fixed cartesian axes and μ_ρ is a space-fixed component of the electric dipole moment. Inserting values for the fundamental constants the relation between G_{ji} and $\boldsymbol{M}_{ji}$ may be expressed in practical units as

$$(G_{ji}/\text{pm}^2) = 41.6238\,|M_{ji}/\text{D}|^2$$

where D (= debye) = $3.335\,641 \times 10^{-30}$ C m.

Net integrated absorption band intensities are usually characterized by one of the quantities A, S, $\bar{S}$, Γ, or G_{net} as defined in the table. The relation between these quantities is given by the (approximate) equations

$$G_{\text{net}} = \Gamma/N_{\text{A}} = A/\tilde{\nu}_0 N_{\text{A}} = S/\tilde{\nu}_0 = \bar{S}(kT/\tilde{\nu}_0)$$

However, only the first equality is exact. The relation to A, $\bar{S}$ and S involves dividing by the band centre wavenumber $\tilde{\nu}_0$ for a band, to correct for the fact that A, $\bar{S}$ and S are obtained by integrating over wavenumber rather than the logarithm of wavenumber used for G_{net} and Γ. This correction is only approximate for a band (although negligible error is involved for single-line intensities in gases). The relation to $\bar{S}$ involves the assumption that the gas is ideal (which is approximately true at low pressures), and also involves the temperature. Thus the quantities Γ and G_{net} are most simply related to more fundamental quantities such as the Einstein transition probabilities and the transition moment, and are the preferred quantities for reporting integrated line or band intensities.

The situation is further complicated by the fact that some authors use the symbol S for any of the above quantities, particularly for any of the quantities here denoted A, S and $\bar{S}$. It is therefore particularly important to define quantities and symbols used in reporting integrated intensities.

For transitions between individual states any of the more fundamental quantities G_{ji}, $B_{\tilde{\nu},ji}$, A_{ji}, or $|M_{ji}|$ may be used; the relations are as given above, and are exact. Note, however, that the integrated absorption coefficient A should not be confused with the Einstein coefficient A_{ji} (nor with absorbance, for which the symbol A is also used). Where such confusion might arise, we recommend writing $\bar{A}$ for the band intensity expressed as an integrated absorption coefficient over wavenumber.

The SI unit and commonly used units of A, S, $\bar{S}$, Γ and G are as in the table below. Also given in the table are numerical conversion factors, using the commonly used units, from A, S, $\bar{S}$, and Γ to G_{net}.

Quantity	*SI unit*	*Common unit*	*Conversion factor*
A, $\bar{A}$	m mol^{-1}	km mol^{-1}	$(G/\text{pm}^2) = 16.605\,40 \dfrac{(A/\text{km mol}^{-1})}{(\tilde{\nu}_0/\text{cm}^{-1})}$
$\bar{S}$	$\text{Pa}^{-1}\,\text{m}^{-2}$	$\text{atm}^{-1}\,\text{cm}^{-2}$	$(G/\text{pm}^2) = 1.362\,603 \times 10^{-2} \dfrac{(\bar{S}/\text{atm}^{-1}\,\text{cm}^{-2})(T/\text{K})}{(\tilde{\nu}_0/\text{cm}^{-1})}$
S	m	cm	$(G/\text{pm}^2) = 10^{20} \dfrac{(S/\text{cm})}{(\tilde{\nu}_0/\text{cm}^{-1})}$
Γ	$\text{m}^2\,\text{mol}^{-1}$	$\text{cm}^2\,\text{mol}^{-1}$	$(G/\text{pm}^2) = 1.660\,540 \times 10^{-4}\,(\Gamma/\text{cm}^2\,\text{mol}^{-1})$
G	m^2	pm^2	

Quantities concerned with spectral absorption intensity and relations among these quantities are discussed in references [59]–[61], and a list of published measurements of line intensities and band intensities for gas phase infrared spectra may be found in references [60] and [61].

2.8 SOLID STATE

The quantities and their symbols given here have been selected from more extensive lists of IUPAP [4] and ISO [5.p]. See also the *International Tables for Crystallography*, Volume A [62].

Name	*Symbol*	*Definition*	*SI unit*	*Notes*
lattice vector	$\boldsymbol{R}, \boldsymbol{R}_0$		m	
fundamental translation vectors for the crystal lattice	$\boldsymbol{a}_1; \boldsymbol{a}_2; \boldsymbol{a}_3$, $\boldsymbol{a}; \boldsymbol{b}; \boldsymbol{c}$	$\boldsymbol{R} = n_1\boldsymbol{a}_1 + n_2\boldsymbol{a}_2 + n_3\boldsymbol{a}_3$	m	1
(circular) reciprocal lattice vector	$\boldsymbol{G}$	$\boldsymbol{G}\cdot\boldsymbol{R} = 2\pi m$	m^{-1}	2
(circular) fundamental translation vectors for the reciprocal lattice	$\boldsymbol{b}_1; \boldsymbol{b}_2; \boldsymbol{b}_3$, $\boldsymbol{a}^*; \boldsymbol{b}^*; \boldsymbol{c}^*$	$\boldsymbol{a}_i\cdot\boldsymbol{b}_k = 2\pi\delta_{ik}$	m^{-1}	3
unit cell lengths	$a; b; c$		m	
unit cell angles	$\alpha; \beta; \gamma$		rad, 1	
reciprocal unit cell lengths	$a^*; b^*; c^*$		m^{-1}	
reciprocal unit cell angles	$\alpha^*; \beta^*; \gamma^*$		rad^{-1}, 1	
fractional coordinates	$x; y; z$	$x = X/a$	1	4
atomic scattering factor	f	$f = E_\text{a}/E_\text{e}$	1	5
structure factor with indices h,k,l	$F(h,k,l)$	$F = \sum_{n=1}^{N} f_n \text{e}^{2\pi\text{i}(hx_n + ky_n + lz_n)}$	1	6
lattice plane spacing	d		m	
Bragg angle	θ	$n\lambda = 2d\sin\theta$	1, rad	
order of reflection	n		1	
order parameters,				
short range	σ		1	
long range	s		1	
Burgers vector	$\boldsymbol{b}$		m	
particle position vector	$\boldsymbol{r}, \boldsymbol{R}_j$		m	7
equilibrium position vector of an ion	$\boldsymbol{R}_0$		m	
displacement vector of an ion	$\boldsymbol{u}$	$\boldsymbol{u} = \boldsymbol{R} - \boldsymbol{R}_0$	m	
Debye–Waller factor	B, D		1	
Debye angular wavenumber	q_D		m^{-1}	
Debye angular frequency	ω_D		s^{-1}	

(1) n_1, n_2 and n_3 are integers. a, b and c are also called the lattice constants.
(2) m is an integer.
(3) Reciprocal lattice vectors are sometimes defined by $\boldsymbol{a}_i\cdot\boldsymbol{b}_k = \delta_{ik}$.
(4) X denotes the coordinate of dimension length.
(5) E_a and E_e denote the scattering amplitudes for the atom and the isolated electron, respectively.
(6) N is the number of atoms in the unit cell.
(7) To distinguish between electron and ion position vectors, lower case and capital letters are used respectively. The subscript j relates to particle j.

Name	*Symbol*	*Definition*	*SI unit*	*Notes*
Grüneisen parameter	γ, Γ	$\gamma = \alpha V/\kappa C_V$	1	8
Madelung constant	$\alpha, \mathscr{M}$	$E_{\text{coul}} = \dfrac{\alpha N_A z_+ z_- e^2}{4\pi\varepsilon_0 R_0}$	1	
density of states	N_E	$N_E = \mathrm{d}N(E)/\mathrm{d}E$	$\mathrm{J^{-1}\,m^{-3}}$	9
(spectral) density of vibrational modes	N_ω, g	$N_\omega = \mathrm{d}N(\omega)/\mathrm{d}\omega$	$\mathrm{s\,m^{-3}}$	10
resistivity tensor	ρ_{ik}	$\boldsymbol{E} = \boldsymbol{\rho}\cdot\boldsymbol{j}$	$\Omega\,\mathrm{m}$	
conductivity tensor	σ_{ik}	$\boldsymbol{\sigma} = \boldsymbol{\rho}^{-1}$	$\mathrm{S\,m^{-1}}$	
thermal conductivity tensor	λ_{ik}	$\boldsymbol{J}_q = -\boldsymbol{\lambda}\cdot\mathrm{grad}\,T$	$\mathrm{W\,m^{-1}\,K^{-1}}$	
residual resistivity	ρ_R		$\Omega\,\mathrm{m}$	
relaxation time	τ	$\tau = l/v_F$	s	11
Lorenz coefficient	L	$L = \lambda/\sigma T$	$\mathrm{V^2\,K^{-2}}$	
Hall coefficient	A_H, R_H	$\boldsymbol{E} = \boldsymbol{\rho}\cdot\boldsymbol{j} + R_H(\boldsymbol{B}\times\boldsymbol{j})$	$\mathrm{m^3\,C^{-1}}$	
thermoelectric force	E		V	12
Peltier coefficient	Π		V	12
Thomson coefficient	$\mu, (\tau)$		$\mathrm{V\,K^{-1}}$	
work function	Φ	$\Phi = E_\infty - E_F$	J	13
number density, number concentration	$n; p$		$\mathrm{m^{-3}}$	14
gap energy	E_g		J	15
donor ionization energy	E_d		J	15
acceptor ionization energy	E_a		J	15
Fermi energy	E_F, ε_F		J	15
circular wave vector, propagation vector	$\boldsymbol{k}; \boldsymbol{q}$	$k = 2\pi/\lambda$	$\mathrm{m^{-1}}$	16
Bloch function	$u_k(\boldsymbol{r})$	$\psi(\boldsymbol{r}) = u_k(\boldsymbol{r})\exp(\mathrm{i}\boldsymbol{k}\cdot\boldsymbol{r})$	$\mathrm{m^{-3/2}}$	17
charge density of electrons	ρ	$\rho(\boldsymbol{r}) = -e\psi^*(\boldsymbol{r})\psi(\boldsymbol{r})$	$\mathrm{C\,m^{-3}}$	17, 18
effective mass	m^*		kg	19
mobility	μ	$\mu = v_{\text{drift}}/E$	$\mathrm{m^2\,V^{-1}\,s^{-1}}$	19
mobility ratio	b	$b = \mu_n/\mu_p$	1	
diffusion coefficient	D	$\mathrm{d}N/\mathrm{d}t = -DA(\mathrm{d}n/\mathrm{d}x)$	$\mathrm{m^2\,s^{-1}}$	19
diffusion length	L	$L = \sqrt{D\tau}$	m	19, 20
characteristic (Weiss) temperature	θ, θ_W		K	
Curie temperature	T_C		K	
Néel temperature	T_N		K	

(8) α is the cubic expansion coefficient, V the volume, κ the isothermal compressibility, and C_V the heat capacity at constant volume.
(9) $N(E)$ is the total number of states of electronic energy less than E, divided by the volume.
(10) $N(\omega)$ is the total number of vibrational modes with circular frequency less than ω, divided by the volume.
(11) The definition applies to electrons in metals; l is the mean free path, and v_F is the electron velocity on the Fermi sphere.
(12) The substances to which the symbol applies are denoted by subscripts.
(13) E_∞ is the electron energy at rest at infinite distance.
(14) Specific number densities are denoted by subscripts: for electrons $n_n, n_-, (n)$; for holes n_p, n_+, p; for donors n_d; for acceptors n_a; for the intrinsic number density n_i ($n_i^2 = n_+ n_-$).
(15) The commonly used unit for this quantity is eV.
(16) $\boldsymbol{k}$ is used for particles, $\boldsymbol{q}$ for phonons.

Symbols for planes and directions in crystals

Miller indices of a crystal face, or of a single net plane	(h, k, l) or (h_1, h_2, h_3)
indices of the Bragg reflection from the set of parallel net planes (h, k, l)	h, k, l or h_1, h_2, h_3
indices of a set of all symmetrically equivalent crystal faces, or net planes	$\{h, k, l\}$ or $\{h_1, h_2, h_3\}$
indices of a lattice direction (zone axis)	$[u, v, w]$
indices of a set of symmetrically equivalent lattice directions	$\langle u, v, w \rangle$

In each of these cases, when the letter symbol is replaced by numbers it is customary to omit the commas. For a single plane or crystal face, or a specific direction, a negative number is indicated by a bar over the number.

Example $(\bar{1}10)$ denotes the parallel planes $h = -1, k = +1, l = 0$.

(i) *Crystal lattice symbols*

primitive	P
face-centred	F
body-centred	I
base-centred	A; B; C
rhombohedral	R

(ii) *Herman–Maugin symbols of symmetry operations*

Operation	*Symbol*	*Examples*
n-fold rotation	n	1; 2; 3; 4; 6
n-fold inversion	$\bar{n}$	$\bar{1}; \bar{2}; \bar{3}; \bar{4}; \bar{6}$
n-fold screw	n_k	$2_1; 3_1; 3_2; \ldots$
reflection	m	
glide	$a; b; c; n; d$	

Notes (continued)

(17) $\psi(\boldsymbol{r})$ is a one-electron wavefunction.

(18) The total charge density is obtained by summing over all electrons.

(19) Subscripts n and p or $-$ and $+$ may be used to denote electrons and holes respectively.

(20) D is the diffusion coefficient and τ the lifetime.

2.9 STATISTICAL THERMODYNAMICS

The names and symbols given here are in agreement with those recommended by IUPAP [4] and by ISO [5.i].

Name	*Symbol*	*Definition*	*SI units*	*Notes*
number of entities	N		1	
number density of entities, number concentration	C, n	$C = N/V$	m^{-3}	
Avogadro constant	L, N_A	$L = N/n$	mol^{-1}	1
Boltzmann constant	k, k_B		$J K^{-1}$	
gas constant (molar)	R	$R = Lk$	$J K^{-1} mol^{-1}$	
molecular position vector	$\boldsymbol{r}(x, y, z)$		m	
molecular velocity vector	$\boldsymbol{c}\ (c_x, c_y, c_z)$, $\boldsymbol{u}\ (u_x, u_y, u_z)$	$\boldsymbol{c} = \mathrm{d}\boldsymbol{r}/\mathrm{d}t$	$m\ s^{-1}$	
molecular momentum vector	$\boldsymbol{p}\ (p_x, p_y, p_z)$	$\boldsymbol{p} = m\boldsymbol{c}$	$kg\ m\ s^{-1}$	
velocity distribution function	$f(c_x)$	$f = \left(\frac{m}{2\pi kT}\right)^{\frac{1}{2}} \exp\left(-\frac{mc_x^2}{2kT}\right)$	$m^{-1}\ s$	
speed distribution function	$F(c)$	$F = 4\pi c^2 \left(\frac{m}{2\pi kT}\right)^{\frac{3}{2}} \exp\left(-\frac{mc^2}{2kT}\right)$	$m^{-1}\ s$	
average speed	$\bar{c}, \bar{u}, \langle c \rangle, \langle u \rangle$	$\bar{c} = \int cF(c)\mathrm{d}c$	$m\ s^{-1}$	
generalized coordinate	q		(m)	2
generalized momentum	p	$p = \partial L/\partial \dot{q}$	$(kg\ m\ s^{-1})$	2
volume in phase space	Ω	$\Omega = (1/h)\int p\,\mathrm{d}q$	1	
probability	P		1	
statistical weight, degeneracy	g, d, W, ω, β		1	3
(cumulative) number of states	N, W		1	
density of states	$\rho(E)$	$\rho(E) = \mathrm{d}N/\mathrm{d}E$	J^{-1}	
partition function, sum over states,				
single molecule	q, z	$q = \sum_i g_i \exp(-\varepsilon_i/kT)$	1	4
canonical ensemble (system, or assembly)	Q, Z		1	
microcanonical ensemble	Ω, z		1	
grand canonical ensemble	Ξ		1	

(1) n is the amount of substance or the chemical amount.
(2) If q is a length then p is a momentum as indicated by the units in parentheses. In the definition of p, L denotes the Lagrangian.
(3) β is usually used for a spin statistical weight.
(4) ε_i denotes the energy of the ith molecular level.

Name	*Symbol*	*Definition*	*SI units*	*Notes*
symmetry number	σ, s		1	
reciprocal temperature parameter	β	$\beta = 1/kT$	J^{-1}	
characteristic temperature	Θ, θ		K	5
absolute activity	λ	$\lambda_B = \exp(\mu_B/RT)$	1	6

(5) Particular characteristic temperatures are denoted with subscripts, e.g. rotational $\Theta_r = hc\tilde{B}/k$, vibrational $\Theta_v = hc\tilde{\nu}/k$, Debye $\Theta_D = hc\tilde{\nu}_D/k$, Einstein $\Theta_E = hc\tilde{\nu}_E/k$.
(6) The definition applies to entities B. μ_B is the chemical potential, see p.49.

2.10 GENERAL CHEMISTRY

The symbols given by IUPAP [4] and by ISO [5.e, i] are in agreement with the recommendations given here.

Name	*Symbol*	*Definition*	*SI unit*	*Notes*
number of entities (e.g. molecules, atoms, ions, formula units)	N		1	
amount (of substance), chemical amount	n	$n_B = N_B/L$	mol	1, 2
Avogadro constant	L, N_A		mol^{-1}	
mass of atom, atomic mass	m_a, m		kg	
mass of entity (molecule, formula unit)	m_f, m		kg	3
atomic mass constant	m_u	$m_u = m_a(^{12}C)/12$	kg	4
molar mass	M	$M_B = m/n_B$	$kg\ mol^{-1}$	2, 5
relative molecular mass, (relative molar mass, molecular weight)	M_r	$M_r = m_f/m_u$	1	6
relative atomic mass, (atomic weight)	A_r	$A_r = m_a/m_u$	1	6
molar volume	V_m	$V_{m,B} = V/n_B$	$m^3\ mol^{-1}$	2, 5
mass fraction	w	$w_j = m_j/\Sigma m_i$	1	7
volume fraction	ϕ	$\phi_j = V_j/\Sigma V_i$	1	7, 8
mole fraction, amount fraction, number fraction	x, y	$x_B = n_B/\Sigma n_A$	1	2, 9

(1) The words 'of substance' may be replaced by the specification of the entity.

Example When the amount of O_2 is equal to 3 moles, $n(O_2) = 3$ mol, then the amount of $\frac{1}{2}O_2$ is equal to 6 moles, $n(\frac{1}{2}O_2) = 6$ mol. Thus $n(\frac{1}{2}O_2) = 2n(O_2)$. See also the discussion on p.46.

(2) The definition applies to entities B which should always be indicated by a subscript or in parentheses, e.g. n_B or $n(B)$.
(3) A formula unit is not a unit but an entity specified as a group of atoms by the way the chemical formula is written. See examples on p.45.
(4) m_u is equal to the unified atomic mass unit, with symbol u, i.e. $m_u = 1$ u (see section 3.7). In biochemistry this unit is called the dalton, with symbol Da, although the name and symbol have not been approved by CGPM.
(5) The definition applies to pure substance, where m is the total mass and V is the total volume. However, corresponding quantities may also be defined for a mixture as m/n and V/n, where $n = \sum_i n_i$. These quantities are called the mean molar mass and the mean molar volume respectively.
(6) For molecules M_r is the relative molecular mass or molecular weight; for atoms M_r is the relative atomic mass or atomic weight and the symbol A_r may be used. M_r may also be called the relative molar mass, $M_{r,B} = M_B/M^{\ominus}$, where $M^{\ominus} = 1\ g\ mol^{-1}$. The standard atomic weights, recommended by IUPAC, are listed in table 6.2, p.94.
(7) The definition applies to component j.
(8) V_j and V_i are the volumes of appropriate components prior to mixing.
(9) For condensed phases x is used, and for gaseous mixtures y may be used.

Name	*Symbol*	*Definition*	*SI unit*	*Notes*
(total) pressure	p, P		Pa	10
partial pressure	p_B	$p_B = y_B p$	Pa	11
mass concentration, (mass density)	γ, ρ	$\gamma_j = m_j/V$	kg m^{-3}	7, 12, 13
number concentration, number density of entities	C, n	$C_B = N_B/V$	m^{-3}	2,12,14
amount concentration, concentration	c	$c_B = n_B/V$	mol m^{-3}	2,12,15
solubility	s	$s_B = c_B$(saturated soln)	mol m^{-3}	2
molality (of a solute)	m, b	$m_B = n_B/m_A$	mol kg^{-1}	2,16
surface concentration	Γ	$\Gamma_B = n_B/A$	mol m^{-2}	2
stoichiometric number	ν		1	17

(10) Pressures are often expressed in the non-SI unit bar, where 1 bar = 10^5 Pa. The standard pressure $p^\ominus$ = 1 bar = 10^5 Pa (see p.54, 112, 166). Low pressures are often expressed in millibars, where 1 mbar = 10^{-3} bar = 100 Pa.

(11) The symbol and the definition apply to molecules B, which should be specified. In real (non-ideal) gases there is a difficulty about defining partial pressure. Some workers regard the equation given as an operational definition; the alternative is to regard the partial pressure of B as the pressure exerted by molecules B.

(12) V is the volume of the mixture.

(13) In polymer science the symbol c is often used for mass concentration.

(14) The term number concentration and symbol C is preferred for mixtures.

(15) The unit mol dm^{-3} is often used for amount concentration. 'Amount concentration' is an abbreviation for 'amount-of-substance concentration'. (The Clinical Chemistry Division of IUPAC recommends that amount of substance concentration be abbreviated to 'substance concentration'.) When there is no risk of confusion the word 'concentration' may be used alone. The symbol [B] is often used for amount concentration of entities B. This quantity is also sometimes called molarity. A solution of, for example, 1 mol dm^{-3} is often called a 1 molar solution, denoted 1 M solution. Thus M is often treated as a symbol for mol dm^{-3}.

(16) In the definition m_B denotes the molality of solute B, and m_A denotes the mass of solvent A; thus the same symbol m is used with two different meanings. This confusion of notation may be avoided by using the symbol b for molality.

A solution of molality 1 mol/kg is occasionally called a 1 molal solution, denoted 1 m solution; however, the symbol m should not be treated as a symbol for the unit mol kg^{-1}.

(17) The stoichiometric number is defined through the reaction equation. It is negative for reactants and positive for products. The values of the stoichiometric numbers depend on how the reaction equation is written.

Example $(1/2)N_2 + (3/2)H_2 = NH_3$: $\nu(N_2) = -1/2$, $\nu(H_2) = -3/2$, $\nu(NH_3) = +1$.

A symbolic way of writing a general chemical equation is

$$0 = \Sigma\, \nu_j B_j$$

where B_j denotes an entity in the reaction. For multireaction systems it is convenient to write the chemical equations in matrix form

$$\boldsymbol{A\nu} = \mathbf{0}$$

where $\boldsymbol{A}$ is the conservation (or formula) matrix with elements A_{ij} representing the number of atoms of the ith element in the jth reaction component (reactant or product) entity and $\boldsymbol{\nu}$ is the stoichiometric number matrix with elements ν_{jk} being the stoichiometric numbers of the jth reaction component entity in the kth reaction. When there are N_s reacting species involved in the system consisting of N_e elements $\boldsymbol{A}$ becomes an $N_e \times N_s$ matrix. Its nullity, $N(\boldsymbol{A}) = N_s - \text{rank}(\boldsymbol{A})$, gives the number of independent chemical reactions, N_r, and the $N_s \times N_r$ stoichiometric number matrix, $\boldsymbol{\nu}$, can be determined as the null space of $\boldsymbol{A}$. $\mathbf{0}$ is an $N_e \times N_r$ zero matrix [63].

Name	*Symbol*	*Definition*	*SI unit*	*Notes*
extent of reaction, advancement	ξ	$n_B = n_{B,0} + \nu_B \xi$	mol	2,18
degree of reaction	α		1	19

(18) $n_{B,0}$ is the amount of B when $\xi = 0$. A more general definition is $\Delta\xi = \Delta n_B/\nu_B$. The extent of reaction also depends on how the reaction equation is written, but it is independent of which entity in the reaction equation is used in the definition.

Example For the reaction in footnote (17), when $\Delta\xi = 2$ mol, $\Delta n(N_2) = -1$ mol, $\Delta n(H_2) = -3$ mol, and $\Delta n(NH_3) = +2$ mol.

This quantity was originally introduced as *degré d'avancement* by de Donder.

(19) For a specific reaction terms such as 'degree of dissociation', 'degree of ionization', etc. are commonly used.

Other symbols and conventions in chemistry

(i) Symbols for particles and nuclear reactions

neutron	n	helion	h
proton	p	alpha particle	α
deuteron	d	electron	e
triton	t	photon	γ
positive muon	μ^+	negative muon	μ^-

The electric charge of particles may be indicated by adding the superscript +, −, or 0; e.g. p^+, n^0, e^-, etc. If the symbols p and e are used without a charge, they refer to the positive proton and negative electron respectively.

The meaning of the symbolic expression indicating a nuclear reaction should be as follows:

$$\text{initial nuclide}\ \left(\text{incoming particles or quanta},\ \text{outgoing particles or quanta}\right)\ \text{final nuclide}$$

Examples $^{14}N(\alpha, p)^{17}O$, $^{59}Co(n, \gamma)^{60}Co$,
$^{23}Na(\gamma, 3n)^{20}Na$, $^{31}P(\gamma, pn)^{29}Si$

(ii) Chemical symbols for the elements

The chemical symbols of elements are (in most cases) derived from their Latin names and consist of one or two letters which should always be printed in roman (upright) type. Only for elements of atomic number greater than 103, the systematic symbols consist of three letters (see footnote U to table 6.2). A complete list is given in table 6.2, p.94. The symbol is not followed by a full stop except at the end of a sentence.

Examples I, U, Pa, C

The symbols can have different meanings:

(a) They can denote an atom of the element. For example, Cl can denote a chlorine atom having 17 protons and 18 or 20 neutrons (giving a mass number of 35 or 37), the difference being ignored. Its mass is on average 35.4527 u in terrestrial samples.

(b) The symbol may, as a kind of shorthand, denote a sample of the element. For example, Fe can denote a sample of iron, and He a sample of helium gas.

The term *nuclide* implies an atom of specified atomic number (proton number) and mass number (nucleon number). Nuclides having the same atomic number but different mass numbers are called isotopic nuclides or *isotopes.* Nuclides having the same mass number but different atomic numbers are called isobaric nuclides or *isobars.*

A nuclide may be specified by attaching the mass number as a left superscript to the symbol for the element. The atomic number may also be attached as a left subscript, if desired, although this is rarely done. If no left superscript is attached, the symbol is read as including all isotopes in natural abundance.

Examples ^{14}N, ^{12}C, ^{13}C, $^{16}_{8}O$, $n(Cl) = n(^{35}Cl) + n(^{37}Cl)$

The ionic charge number is denoted by a right superscript, or by the sign alone when the charge is equal to one.

Examples	Na^{+}	a sodium positive ion (cation)
	$^{79}Br^{-}$	a bromine-79 negative ion (anion, bromide ion)
	Al^{3+} or Al^{+3}	aluminium triply positive ion
	$3\,S^{2-}$ or $3\,S^{-2}$	three sulfur doubly negative ions (sulfide ions)

The right superscript position is also used to convey other information: excited electronic states may be denoted by an asterisk.

Examples H^{*}, Cl^{*}

Oxidation numbers are denoted by positive or negative roman numerals or by zero (see also (iv) below).

Examples Mn^{VII}, O^{-II}, Ni^{0}

The positions and meanings of indices around the symbol of the element are summarized as follows:

left superscript	mass number
left subscript	atomic number
right superscript	charge number, oxidation number, excitation symbol
right subscript	number of atoms per entity (see (iii) below)

(iii) Chemical formulae

Chemical formulae denote entities composed of more than one atom (molecules, complex ions, groups of atoms, etc.).

Examples N_2, P_4, C_6H_6, $CaSO_4$, $PtCl_4{}^{2-}$, $Fe_{0.91}S$

They may also be used as a shorthand to denote a sample of the corresponding chemical substance.

Examples	CH_3OH	methanol
	$\rho(H_2SO_4)$	mass density of sulfuric acid

The number of atoms in an entity is indicated by a right subscript (the numeral 1 being omitted). Groups of atoms may also be enclosed in parentheses. Entities may be specified by giving the corresponding formula, often multiplied by a factor. Charge numbers of complex ions, and excitation symbols, are added as right superscripts to the whole formula. The free radical nature of some entities may be stressed by adding a dot to the symbol.

Examples	H_2O	one water molecule, water
	$\frac{1}{2}O_2$	half an oxygen molecule
	$Zn_3(PO_4)_2$	one zinc phosphate formula unit, zinc phosphate
	$2MgSO_4$	two formula units of magnesium sulfate
	$\frac{1}{5}KMnO_4$	one-fifth of a potassium permanganate formula unit
	$\frac{1}{2}SO_4^{2-}$	half a sulfate ion
	$(CH_3)^{\bullet}$	methyl free radical
	$CH_3\dot{C}HCH_3$	isopropyl radical
	$NO_2{}^*$	electronically excited nitrogen dioxide molecule

In the above examples, $\frac{1}{2}O_2$, $\frac{1}{5}KMnO_4$ and $\frac{1}{2}SO_4^{2-}$ are artificial in the sense that such fractions of a molecule cannot exist. However, it may often be convenient to specify entities in this way when calculating amounts of substance; see (v) below.

Specific electronic states of entities (atoms, molecules, ions) can be denoted by giving the electronic term symbol (see section 2.6) in parentheses. Vibrational and rotational states can be specified by giving the corresponding quantum numbers.

Examples	$Hg(^3P_1)$	a mercury atom in the triplet-P-one state
	$HF(v = 2, J = 6)$	a hydrogen fluoride molecule in the vibrational state $v = 2$ and the rotational state $J = 6$
	$H_2O^+(^2A_1)$	a water molecule ion in the doublet-A-one state

Chemical formulae may be written in different ways according to the information that they convey, as follows:

Formula	*Information conveyed*	*Example for lactic acid*
empirical	stoichiometric proportion only	CH_2O
molecular	in accord with molecular mass	$C_3H_6O_3$
structural	structural arrangement of atoms	$CH_3CHOHCOOH$
displayed	projection of atoms and bonds	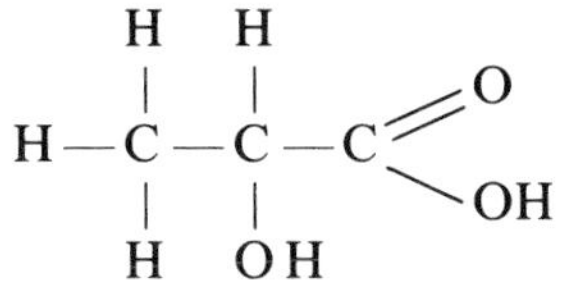
stereochemical	stereochemical arrangement	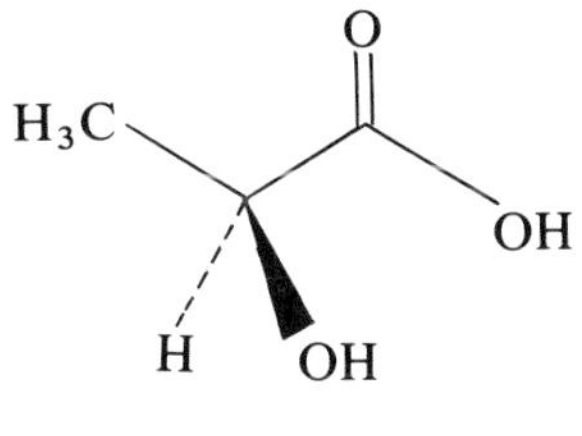

Further conventions for writing chemical formulae are described in [22, 23].

(iv) Equations for chemical reactions

Symbols connecting the reactants and products in a chemical reaction equation have the following meanings:

$H_2 + Br_2 = 2HBr$	stoichiometric relation
$H_2 + Br_2 \rightarrow 2HBr$	net forward reaction
$H_2 + Br_2 \leftrightarrows 2HBr$	reaction, both directions
$H_2 + Br_2 \rightleftharpoons 2HBr$	equilibrium

A single arrow is also used to designate an elementary reaction, such as $H^{\cdot} + Br_2 \rightarrow HBr + Br^{\cdot}$. It should therefore be made clear if this is the usage intended.

Redox equations are often written so that the absolute value of the stoichiometric number for the electrons transferred (which are normally omitted from the overall equation) is equal to one.

Example $(1/5)\,KMn^{VII}O_4 + (8/5)\,HCl = (1/5)\,Mn^{II}Cl_2 + (1/2)\,Cl_2 + (1/5)\,KCl + (4/5)\,H_2O$

Similarly a reaction in an electrochemical cell may be written so that the charge number of the cell reaction is equal to one:

Example $(1/3)\,In^0(s) + (1/2)Hg^I{}_2SO_4(s) = (1/6)In^{III}{}_2(SO_4)_3(aq) + Hg^0(l)$

(the symbols in parentheses denote the state; see (vi) below).

(v) Amount of substance and the specification of entities

The quantity 'amount of substance' or 'chemical amount' ('Stoffmenge' in German) has been used by chemists for a long time without a proper name. It was simply referred to as the 'number of moles'. This practice should be abandoned, because it is wrong to confuse the name of a physical quantity with the name of a unit (in a similar way it would be wrong to use 'number of metres' as a synonym for 'length'). The amount of substance is proportional to the number of specified elementary entities of that substance; the proportionality factor is the same for all substances and is the reciprocal of the Avogadro constant. The elementary entities may be chosen as convenient, not necessarily as physically real individual particles. Since the amount of substance and all physical quantities derived from it depend on this choice it is essential to specify the entities to avoid ambiguities.

Examples	n_{Cl}, $n(Cl)$	amount of Cl, amount of chlorine atoms
	$n(Cl_2)$	amount of Cl_2, amount of chlorine molecules
	$n(H_2SO_4)$	amount of (entities) H_2SO_4
	$n(\frac{1}{5}KMnO_4)$	amount of (entities) $\frac{1}{5}KMnO_4$
	$M(P_4)$	molar mass of (tetraphosphorus) P_4
	c_{HCl}, $c(HCl)$, $[HCl]$	amount concentration of HCl
	$\Lambda(MgSO_4)$	molar conductivity of (magnesium sulfate entities) $MgSO_4$
	$\Lambda(\frac{1}{2}MgSO_4)$	molar conductivity of (entities) $\frac{1}{2}MgSO_4$
	$n(\frac{1}{5}KMnO_4) = 5n(KMnO_4)$	
	$\lambda(\frac{1}{2}Mg^{2+}) = \frac{1}{2}\lambda(Mg^{2+})$	
	$[\frac{1}{2}H_2SO_4] = 2[H_2SO_4]$	
	(See also examples in section 3.2, p.70.)	

Note that 'amount of sulfur' is an ambiguous statement, because it might imply $n(S)$, $n(S_8)$, or $n(S_2)$, etc. In some cases analogous statements are less ambiguous. Thus for compounds the implied entity is usually the molecule or the common formula entity, and for solid metals it is the atom.

Examples '2 moles of water' implies $n(H_2O) = 2$ mol; '0.5 moles of sodium chloride' implies $n(NaCl) = 0.5$ mol; '3 millimoles of iron' implies $n(Fe) = 3$ mmol, but such statements should be avoided whenever there might be ambiguity.

However, in the equation $pV = nRT$ and in equations involving colligative properties, the entity implied in the definition of n should be an individually translating particle (a whole molecule for a gas), whose nature is unimportant.

(vi) States of aggregation

The following one-, two- or three-letter symbols are used to represent the states of aggregation of chemical species [1.j]. The letters are appended to the formula symbol in parentheses, and should be printed in roman (upright) type without a full stop (period).

g	gas or vapour	vit	vitreous substance
l	liquid	a, ads	species adsorbed on a substrate
s	solid	mon	monomeric form
cd	condensed phase (i.e. solid or liquid)	pol	polymeric form
		sln	solution
fl	fluid phase (i.e. gas or liquid)	aq	aqueous solution
		aq, ∞	aqueous solution at infinite dilution
cr	crystalline		
lc	liquid crystal	am	amorphous solid

Examples

$HCl(g)$	hydrogen chloride in the gaseous state
$C_V(\mathrm{fl})$	heat capacity of a fluid at constant volume
$V_\mathrm{m}(\mathrm{lc})$	molar volume of a liquid crystal
$U(\mathrm{cr})$	internal energy of a crystalline solid
$MnO_2(\mathrm{am})$	manganese dioxide as an amorphous solid
$MnO_2(\mathrm{cr, I})$	manganese dioxide as crystal form I
$NaOH(\mathrm{aq})$	aqueous solution of sodium hydroxide
$NaOH(\mathrm{aq}, \infty)$	. . . as above, at infinite dilution
$\Delta_\mathrm{f}H^\ominus(H_2O, \mathrm{l})$	standard enthalpy of formation of liquid water

The symbols g, l, to denote gas phase, liquid phase, etc., are also sometimes used as a right superscript, and the Greek letter symbols α, β, may be similarly used to denote phase α, phase β, etc., in a general notation.

Examples $V_\mathrm{m}^{\mathrm{l}}$, $V_\mathrm{m}^{\mathrm{s}}$ molar volume of the liquid phase, ... of the solid phase
S_m^{α}, S_m^{β} molar entropy of phase α, ... of phase β

2.11 CHEMICAL THERMODYNAMICS

The names and symbols of the more generally used quantities given here are also recommended by IUPAP [4] and by ISO [5.e, i]. Additional information can be found in [1.d, j and 24]

Name	*Symbol*	*Definition*	*SI unit*	*Notes*
heat	q, Q		J	1
work	w, W		J	1
internal energy	U	$\Delta U = q + w$	J	1
enthalpy	H	$H = U + pV$	J	
thermodynamic temperature	T		K	
Celsius temperature	θ, t	$\theta/°\mathrm{C} = T/\mathrm{K} - 273.15$	°C	2
entropy	S	$\mathrm{d}S = \mathrm{d}q_{\mathrm{rev}}/T$	$\mathrm{J\,K^{-1}}$	
Helmholtz energy, (Helmholtz function)	A	$A = U - TS$	J	3
Gibbs energy, (Gibbs function)	G	$G = H - TS$	J	
Massieu function	J	$J = -A/T$	$\mathrm{J\,K^{-1}}$	
Planck function	Y	$Y = -G/T$	$\mathrm{J\,K^{-1}}$	
surface tension	γ, σ	$\gamma = (\partial G/\partial A_{\mathrm{s}})_{T,p}$	$\mathrm{J\,m^{-2}}$, $\mathrm{N\,m^{-1}}$	
molar quantity X	$X_{\mathrm{m}}, (\bar{X})$	$X_{\mathrm{m}} = X/n$	(varies)	4, 5
specific quantity X	x	$x = X/m$	(varies)	4, 5
pressure coefficient	β	$\beta = (\partial p/\partial T)_V$	$\mathrm{Pa\,K^{-1}}$	
relative pressure coefficient	α_p	$\alpha_p = (1/p)(\partial p/\partial T)_V$	$\mathrm{K^{-1}}$	
compressibility,				
isothermal	κ_T	$\kappa_T = -(1/V)(\partial V/\partial p)_T$	$\mathrm{Pa^{-1}}$	
isentropic	κ_S	$\kappa_S = -(1/V)(\partial V/\partial p)_S$	$\mathrm{Pa^{-1}}$	
linear expansion coefficient	α_l	$\alpha_l = (1/l)(\partial l/\partial T)$	$\mathrm{K^{-1}}$	
cubic expansion coefficient	α, α_V, γ	$\alpha = (1/V)(\partial V/\partial T)_p$	$\mathrm{K^{-1}}$	6
heat capacity,				
at constant pressure	C_p	$C_p = (\partial H/\partial T)_p$	$\mathrm{J\,K^{-1}}$	
at constant volume	C_V	$C_V = (\partial U/\partial T)_V$	$\mathrm{J\,K^{-1}}$	
ratio of heat capacities	$\gamma, (\kappa)$	$\gamma = C_p/C_V$	1	
Joule–Thomson coefficient	μ, μ_{JT}	$\mu = (\partial T/\partial p)_H$	$\mathrm{K\,Pa^{-1}}$	

(1) Both $q > 0$ and $w > 0$ indicate an increase in the energy of the system; $\Delta U = q + w$. The given equation is sometimes written as $\mathrm{d}U = đq + đw$, where đ denotes an inexact differential.
(2) This quantity is sometimes misnamed 'centigrade temperature'.
(3) It is sometimes convenient to use the symbol F for Helmholtz energy in the context of surface chemistry, to avoid confusion with A for area.
(4) The definition applies to pure substance. However, the concept of molar and specific quantities (see section 1.4, p.7) may also be applied to mixtures.
(5) X is an extensive quantity. The unit depends on the quantity. In the case of molar quantities the entities should be specified.

Example molar volume of B, $V_{\mathrm{m}}(\mathrm{B}) = V/n_{\mathrm{B}}$

(6) This quantity is also called the coefficient of thermal expansion, or the expansivity coefficient.

Name	Symbol	Definition	SI unit	Notes
virial coefficient,				
second	B	$pV_m = RT(1 + B/V_m$	$m^3\ mol^{-1}$	
third	C	$+ C/V_m^2 + \ldots)$	$m^6\ mol^{-2}$	
van der Waals	a	$(p + a/V_m^2)(V_m - b) = RT$	$J\ m^3\ mol^{-2}$	7
coefficients	b		$m^3\ mol^{-1}$	7
compression factor, (compressibility factor)	Z	$Z = pV_m/RT$	1	
partial molar quantity X	X_B, $(\bar{X}_B)$	$X_B = (\partial X/\partial n_B)_{T,p,n_{j \neq B}}$	(varies)	8
chemical potential, (partial molar Gibbs energy)	μ	$\mu_B = (\partial G/\partial n_B)_{T,p,n_{j \neq B}}$	$J\ mol^{-1}$	9
standard chemical potential	$\mu^\ominus$, μ°		$J\ mol^{-1}$	10
absolute activity	λ	$\lambda_B = \exp(\mu_B/RT)$	1	9
(relative) activity	a	$a_B = \exp\left[\dfrac{\mu_B - \mu_B^\ominus}{RT}\right]$	1	9,11
standard partial molar enthalpy	$H_B^\ominus$	$H_B^\ominus = \mu_B^\ominus + TS_B^\ominus$	$J\ mol^{-1}$	9,10
standard partial molar entropy	$S_B^\ominus$	$S_B^\ominus = -(\partial \mu_B^\ominus/\partial T)_p$	$J\ mol^{-1}\ K^{-1}$	9,10
standard reaction Gibbs energy (function)	$\Delta_r G^\ominus$	$\Delta_r G^\ominus = \sum_B \nu_B \mu_B^\ominus$	$J\ mol^{-1}$	10,12, 13,14
affinity of reaction	A, $(\mathcal{A})$	$A = -(\partial G/\partial \xi)_{p,T} = -\sum_B \nu_B \mu_B$	$J\ mol^{-1}$	13

(7) For a gas satisfying the van der Waals equation of state, given in the definition, the second virial coefficient is related to the parameters a and b in the van der Waals equation by

$$B = b - a/RT$$

(8) The symbol applies to entities B which should be specified. The bar may be used to distinguish partial molar X from X when necessary.

Example The partial molar volume of Na_2SO_4 in aqueous solution may be denoted $\bar{V}(Na_2SO_4, aq)$, in order to distinguish it from the volume of the solution $V(Na_2SO_4, aq)$.

(9) The definition applies to entities B which should be specified.
(10) The symbol $^\ominus$ or $^\circ$ is used to indicate standard. They are equally acceptable. Definitions of standard states are discussed below (p.53). Whenever a standard chemical potential $\mu^\ominus$ or a standard equilibrium constant $K^\ominus$ or other standard quantity is used, the standard state must be specified.
(11) In the defining equation given here the pressure dependence of the activity has been neglected as is often done for condensed phases at atmospheric pressure.
An equivalent definition is $a_B = \lambda_B/\lambda_B^\ominus$, where $\lambda_B^\ominus = \exp(\mu_B^\ominus/RT)$. The definition of $\mu^\ominus$ depends on the choice of the standard state; see Section (iv) on p.53.
(12) The symbol r indicates reaction in general. In particular cases r can be replaced by another appropriate subscript, e.g. $\Delta_f H^\ominus$ denotes the standard molar enthalpy of formation; see p.51 below for a list of subscripts.
(13) The reaction must be specified for which this quantity applies.
(14) Reaction enthalpies (and reaction energies in general) are usually quoted in $kJ\ mol^{-1}$. In older literature $kcal\ mol^{-1}$ is also common, where 1 kcal = 4.184 kJ (see p.112).

Name	*Symbol*	*Definition*	*SI unit*	*Notes*
standard reaction enthalpy	$\Delta_r H^\ominus$	$\Delta_r H^\ominus = \sum_B \nu_B H_B{}^\ominus$	$J\ mol^{-1}$	10,12, 13,14
standard reaction entropy	$\Delta_r S^\ominus$	$\Delta_r S^\ominus = \sum_B \nu_B S_B{}^\ominus$	$J\ mol^{-1}\ K^{-1}$	10,12,13
reaction quotient	Q	$Q = \prod_B a_B{}^{\nu_B}$	1	15
equilibrium constant	$K^\ominus$, K	$K^\ominus = \exp(-\Delta_r G^\ominus/RT)$	1	10,13,16
equilibrium constant,				
pressure basis	K_p	$K_p = \prod_B p_B{}^{\nu_B}$	$Pa^{\Sigma\nu}$	13,17
concentration basis	K_c	$K_c = \prod_B c_B{}^{\nu_B}$	$(mol\ m^{-3})^{\Sigma\nu}$	13,17
molality basis	K_m	$K_m = \prod_B m_B{}^{\nu_B}$	$(mol\ kg^{-1})^{\Sigma\nu}$	13,17
fugacity	f, $\tilde{p}$	$f_B = \lambda_B \lim_{p\to 0} (p_B/\lambda_B)_T$	Pa	9
fugacity coefficient	ϕ	$\phi_B = f_B/p_B$	1	
Henry's law constant	k_H	$k_{H,B} = \lim_{x_B\to 0} (f_B/x_B) = (\partial f_B/\partial x_B)_{x_B=0}$	Pa	9, 18
activity coefficient				
referenced to Raoult's law	f	$f_B = a_B/x_B$	1	9,19
referenced to Henry's law				
molality basis	γ_m	$a_{m,B} = \gamma_{m,B} m_B/m^\ominus$	1	9, 20
concentration basis	γ_c	$a_{c,B} = \gamma_{c,B} c_B/c^\ominus$	1	9, 20
mole fraction basis	γ_x	$a_{x,B} = \gamma_{x,B} x_B$	1	9, 20
ionic strength,				
molality basis	I_m, I	$I_m = \frac{1}{2}\Sigma m_B z_B{}^2$	$mol\ kg^{-1}$	
concentration basis	I_c, I	$I_c = \frac{1}{2}\Sigma c_B z_B{}^2$	$mol\ m^{-3}$	

(15) This quantity applies in general to a system which is not in equilibrium.
(16) This quantity is equal to the value of Q in equilibrium, when the affinity is zero. It is dimensionless and its value depends on the choice of standard state, which must be specified. ISO [5.i] and the IUPAC Thermodynamics Commission [24] recommend the symbol $K^\ominus$ and the name 'standard equilibrium constant', but some thermodynamicists prefer the symbol K and the name 'thermodynamic equilibrium constant'.
(17) These quantities are not in general dimensionless. One can define in an analogous way an equilibrium constant in terms of fugacity K_f, etc. At low pressures K_p is approximately related to $K^\ominus$ by the equation $K^\ominus \approx K_p/(p^\ominus)^{\Sigma\nu}$, and similarly in dilute solutions K_c is approximately related to $K^\ominus$ by $K^\ominus \approx K_c/(c^\ominus)^{\Sigma\nu}$; however, the exact relations involve fugacity coefficients or activity coefficients [24].

The equilibrium constant of dissolution of an electrolyte (describing the equilibrium between excess solid phase and solvated ions) is often called a solubility product, denoted K_{sol} or K_s (or $K_{sol}^\ominus$ or $K_s^\ominus$ as appropriate). In a similar way the equilibrium constant for an acid dissociation is often written K_a, for base hydrolysis K_b, and for water dissociation K_w.
(18) Henry's law is sometimes expressed in terms of molalities or concentrations and then the corresponding units of the Henry's law constant are $Pa\ kg\ mol^{-1}$ or $Pa\ m^3\ mol^{-1}$, respectively.
(19) This quantity applies to pure phases, substances in mixtures, or solvents.
(20) This quantity applies to solutes.

Name	*Symbol*	*Definition*	*SI unit*	*Notes*
osmotic coefficient,				
molality basis	ϕ_m	$\phi_m = \dfrac{\mu_A{}^* - \mu_A}{RTM_A \Sigma m_B}$	1	
mole fraction basis	ϕ_x	$\phi_x = \dfrac{\mu_A - \mu_A{}^*}{RT \ln x_A}$	1	
osmotic pressure	Π	$\Pi = c_B RT$	Pa	21

(21) The defining equation applies to ideal dilute solutions. The entities B are individually moving solute molecules, ions, etc. regardless of their nature. Their amount is sometimes expressed in osmoles (meaning a mole of osmotically active entities), but this use is discouraged.

Other symbols and conventions in chemical thermodynamics

A more extensive description of this subject can be found in [24].

(i) Symbols used as subscripts to denote a chemical process or reaction

These symbols should be printed in roman (upright) type, without a full stop (period).

vaporization, evaporation (liquid → gas)	vap
sublimation (solid → gas)	sub
melting, fusion (solid → liquid)	fus
transition (between two phases)	trs
mixing of fluids	mix
solution (of solute in solvent)	sol
dilution (of a solution)	dil
adsorption	ads
displacement	dpl
immersion	imm
reaction in general	r
atomization	at
combustion reaction	c
formation reaction	f

(ii) Recommended superscripts

standard	⦵, ○
pure substance	*
infinite dilution	∞
ideal	id
activated complex, transition state	‡
excess quantity	E

(iii) Examples of the use of these symbols

The subscripts used to denote a chemical process, listed under (i) above, should be used as subscripts to the Δ symbol to denote the change in an extensive thermodynamic quantity associated with the process.

Example $\Delta_{\mathrm{vap}} H = H(\mathrm{g}) - H(\mathrm{l})$, for the enthalpy of vaporization, an extensive quantity proportional to the amount of substance vaporized.

The more useful quantity is usually the change divided by the amount of substance transferred, which should be denoted with an additional subscript m.

Example $\Delta_{\mathrm{vap}} H_{\mathrm{m}}$ for the molar enthalpy of vaporization.

However, the subscript m is frequently omitted, particularly when the reader may tell from the units that a molar quantity is implied.

Example $\Delta_{\mathrm{vap}} H = 40.7\ \mathrm{kJ\,mol^{-1}}$ for H_2O at 373.15 K and 1 atm.

The subscript specifying the change is also sometimes attached to the symbol for the quantity rather than the Δ, so that the above quantity is denoted $\Delta H_{\mathrm{vap,m}}$ or simply ΔH_{vap}, but this is not recommended.

The subscript r is used to denote changes associated with a *chemical reaction*. Although symbols such as $\Delta_{\mathrm{r}} H$ should denote the integral enthalpy of reaction, $\Delta_{\mathrm{r}} H = H(\xi_2) - H(\xi_1)$, in practice this symbol is usually used to denote the change divided by the amount transferred, i.e. the change per extent of reaction, defined by the equation

$$\Delta_{\mathrm{r}} H = \sum_{\mathrm{B}} \nu_{\mathrm{B}} H_{\mathrm{B}} = (\partial H / \partial \xi)_{T,p}$$

It is thus essential to specify the stoichiometric reaction equation when giving numerical values for such quantities in order to define the extent of reaction ξ and the values of the stoichiometric numbers ν_{B}.

Example $N_2(\mathrm{g}) + 3H_2(\mathrm{g}) = 2NH_3(\mathrm{g})$, $\Delta_{\mathrm{r}} H^\ominus = -92.4\ \mathrm{kJ\,mol^{-1}}$
$\Delta_{\mathrm{r}} S^\ominus = -199\ \mathrm{J\,mol^{-1}\,K^{-1}}$

The $\mathrm{mol^{-1}}$ in the units identifies the quantities in this example as the change per extent of reaction. They may be called the molar enthalpy and entropy of reaction, and a subscript m may be added to the symbol, to emphasize the difference from the integral quantities if required.

The *standard reaction quantities* are particularly important. They are defined by the equations

$$\Delta_{\mathrm{r}} H^\ominus \ (= \Delta_{\mathrm{r}} H_{\mathrm{m}}{}^\ominus = \Delta H_{\mathrm{m}}{}^\ominus) = \sum_{\mathrm{B}} \nu_{\mathrm{B}} H_{\mathrm{B}}{}^\ominus$$

$$\Delta_{\mathrm{r}} S^\ominus \ (= \Delta_{\mathrm{r}} S_{\mathrm{m}}{}^\ominus = \Delta S_{\mathrm{m}}{}^\ominus) = \sum_{\mathrm{B}} \nu_{\mathrm{B}} S_{\mathrm{B}}{}^\ominus$$

$$\Delta_{\mathrm{r}} G^\ominus \ (= \Delta_{\mathrm{r}} G_{\mathrm{m}}{}^\ominus = \Delta G_{\mathrm{m}}{}^\ominus) = \sum_{\mathrm{B}} \nu_{\mathrm{B}} \mu_{\mathrm{B}}{}^\ominus$$

The symbols in parentheses are alternatives. In view of the variety of styles in current use it is important to specify notation with care for these symbols. The relation to the affinity of the reaction is

$$-A = \Delta_{\mathrm{r}} G = \Delta_{\mathrm{r}} G^\ominus + RT \ln \left(\prod_{\mathrm{B}} a_{\mathrm{B}}{}^{\nu_{\mathrm{B}}} \right),$$

and the relation to the standard equilibrium constant is $\Delta_{\mathrm{r}} G^\ominus = -RT \ln K^\ominus$.

The term *combustion* and symbol c denote the complete oxidation of a substance. For the definition of complete oxidation of substances containing elements other than C, H and O see [64]. The corresponding reaction equation is written so that the stoichiometric number ν of the substance is -1.

Example The standard enthalpy of combustion of gaseous methane is $\Delta_{\mathrm{c}} H^\ominus (CH_4, \mathrm{g}, 298.15\ \mathrm{K}) = -890.3\ \mathrm{kJ\,mol^{-1}}$, implying the reaction $CH_4(\mathrm{g}) + 2O_2(\mathrm{g}) \rightarrow CO_2(\mathrm{g}) + 2H_2O(\mathrm{l})$.

The term *formation* and symbol f denote the formation of the substance from elements in their reference state (usually the most stable state of each element at the chosen temperature and standard pressure). The corresponding reaction equation is written so that the stoichiometric number ν of the substance is $+1$.

Example The standard entropy of formation of crystalline mercury II chloride is $\Delta_f S^\circ$ ($HgCl_2$, cr, 298.15 K) $= -154.3\ \text{J mol}^{-1}\ \text{K}^{-1}$, implying the reaction $Hg(l) + Cl_2(g) \rightarrow HgCl_2(cr)$.

The term *atomization*, symbol at, denotes a process in which a substance is separated into its constituent atoms in the ground state in the gas phase. The corresponding reaction equation is written so that the stoichiometric number ν of the substance is -1.

Example The standard (internal) energy of atomization of liquid water is $\Delta_{at} U^\circ (H_2O, l) = 625\ \text{kJ mol}^{-1}$, implying the reaction $H_2O(l) \rightarrow 2H(g) + O(g)$.

(iv) Standard states [1.j, 24]

The standard chemical potential of substance B at temperature T, $\mu_B{}^\circ(T)$, is the value of the chemical potential under standard conditions, specified as follows. Three differently defined standard states are recognized.

For a gas phase. The standard state for a gaseous substance, whether pure or in a gaseous mixture, is the (hypothetical) state of the pure substance B in the gaseous phase at the standard pressure $p = p^\circ$ and exhibiting ideal gas behaviour. The standard chemical potential is defined as

$$\mu_B{}^\circ(T) = \lim_{p \to 0} \left[\mu_B(T, p, y_B, \ldots) - RT \ln(y_B p / p^\circ)\right]$$

For a pure phase, or a mixture, or a solvent, in the liquid or solid state. The standard state for a liquid or solid substance, whether pure or in a mixture, or for a solvent, is the state of the pure substance B in the liquid or solid phase at the standard pressure $p = p^\circ$. The standard chemical potential is defined as

$$\mu_B{}^\circ(T) = \mu_B{}^*(T, p^\circ)$$

For a solute in solution. For a solute in a liquid or solid solution the standard state is referenced to the ideal dilute behaviour of the solute. It is the (hypothetical) state of solute B at the standard molality m°, standard pressure p°, and exhibiting infinitely diluted solution behaviour. The standard chemical potential is defined as

$$\mu_B{}^\circ(T) = [\mu_B(T, p^\circ, m_B, \ldots) - RT \ln(m_B/m^\circ)]^\infty.$$

The chemical potential of the solute B as a function of the molality m_B at constant pressure $p = p^\circ$ is then given by the expression

$$\mu_B(m_B) = \mu_B{}^\circ + RT \ln(m_B \gamma_{m,B}/m^\circ)$$

Sometimes (amount) concentration c is used as a variable in place of molality m; both of the above equations then have c in place of m throughout. Occasionally mole fraction x is used in place of m; both of the above equations then have x in place of m throughout, and $x^\circ = 1$. Although the standard state of a solute is always referenced to ideal dilute behaviour, the definition of the standard state and the value of the standard chemical potential μ° are different depending on whether molality m, concentration c, or mole fraction x is used as a variable.

(v) Standard pressures, molality, and concentration

In principle one may choose any values for the standard pressure $p^{\ominus}$, the standard molality $m^{\ominus}$, and the standard concentration $c^{\ominus}$, although the choice must be specified. For example, in tabulating data appropriate to high pressure chemistry it may be convenient to choose a value of $p^{\ominus} = 1$ kbar.

In practice, however, the most common choice is

$$p^{\ominus} = 10^5 \text{ Pa } (= 1 \text{ bar})$$
$$m^{\ominus} = 1 \text{ mol kg}^{-1}$$
$$c^{\ominus} = 1 \text{ mol dm}^{-3}$$

These values for $m^{\ominus}$ and $c^{\ominus}$ are universally accepted. The value for $p^{\ominus}$, 10^5 Pa, is the IUPAC recommendation since 1982 [1.j], and is recommended for tabulating thermodynamic data. Prior to 1982 the standard pressure was usually taken to be $p^{\ominus} = 101\,325$ Pa (= 1 atm, called the *standard atmosphere*). In any case, the value for $p^{\ominus}$ should be specified.

The conversion of values corresponding to different $p^{\ominus}$ is described in [65]. The newer value of $p^{\ominus} = 10^5$ Pa is sometimes called the *standard state pressure*.

(vi) Thermodynamic properties

Values of many thermodynamic quantities represent basic chemical properties of substances and serve for further calculations. Extensive tabulations exist, e.g. [66–68]. Special care has to be taken in reporting the data and their uncertainties [25, 26].

2.12 CHEMICAL KINETICS

The recommendations given here are based on previous IUPAC recommendations [1.c, k and 27], which are not in complete agreement. Recommendations regarding photochemistry are given in [28] and for recommendations on reporting of chemical kinetics data see also [69].

Name	*Symbol*	*Definition*	*SI unit*	*Notes*
rate of change of quantity X	$\dot{X}$	$\dot{X} = \mathrm{d}X/\mathrm{d}t$	(varies)	1
rate of conversion	$\dot{\xi}$	$\dot{\xi} = \mathrm{d}\xi/\mathrm{d}t$	$\mathrm{mol\ s^{-1}}$	2
rate of concentration change (due to chemical reaction)	r_B, v_B	$r_B = \mathrm{d}c_B/\mathrm{d}t$	$\mathrm{mol\ m^{-3}\ s^{-1}}$	3, 4
rate of reaction (based on amount concentration)	v	$v = \dot{\xi}/V$ $= \nu_B^{-1}\,\mathrm{d}c_B/\mathrm{d}t$	$\mathrm{mol\ m^{-3}\ s^{-1}}$	2, 4
partial order of reaction	n_B, m_B	$v = k\prod c_B^{n_B}$	1	5
overall order of reaction	n, m	$n = \sum n_B$	1	
rate constant, rate coefficient	k	$v = k\prod c_B^{n_B}$	$(\mathrm{m^3\ mol^{-1}})^{n-1}\ \mathrm{s^{-1}}$	6
Boltzmann constant	k, k_B		$\mathrm{J\ K^{-1}}$	
half life	$t_{\frac{1}{2}}$	$c(t_{\frac{1}{2}}) = c(0)/2$	s	
relaxation time	τ		s	7
(Arrhenius) activation energy	E_a, E_A	$E_a = RT^2\,\mathrm{d}\ln k/\mathrm{d}T$	$\mathrm{J\ mol^{-1}}$	8

(1) E.g. rate of change of pressure $\dot{p} = \mathrm{d}p/\mathrm{d}t$, for which the SI unit is $\mathrm{Pa\ s^{-1}}$.
(2) The reaction must be specified for which this quantity applies.
(3) The symbol and the definition apply to entities B.
(4) Note that r_B and v can also be defined on the basis of partial pressure, number concentration, surface concentration, etc., with analogous definitions. If necessary differently defined rates of reaction can be distinguished by a subscript, e.g. $v_p = \nu_B^{-1}\,\mathrm{d}p_B/\mathrm{d}t$, etc. Note that the rate of reaction can only be defined for a reaction of known and time-independent stoichiometry, in terms of a specified reaction equation; also the second equation for the rate of reaction follows from the first only if the volume V is constant. The derivatives must be those due to the chemical reaction considered; in open systems, such as flow systems, effects due to input and output processes must also be taken into account.
(5) The symbol applies to reactant B. The symbol m may be used when confusion with n for amount of substance occurs.
(6) Rate constants k and pre-exponential factors A are usually quoted in either $(\mathrm{dm^3\ mol^{-1}})^{n-1}\ \mathrm{s^{-1}}$ or on a molecular scale in $(\mathrm{cm^3})^{n-1}\ \mathrm{s^{-1}}$ or $(\mathrm{cm^3\ molecule^{-1}})^{n-1}\ \mathrm{s^{-1}}$. Note that 'molecule' is not a unit, but is often included for clarity. Rate constants are frequently quoted as decadic logarithms.

Example For a second order reaction $k = 10^{8.2}\ \mathrm{dm^3\ mol^{-1}\ s^{-1}}$ or $\lg(k/\mathrm{dm^3\ mol^{-1}\ s^{-1}}) = 8.2$
or alternatively $k = 10^{-12.6}\ \mathrm{cm^3\ s^{-1}}$ or $\lg(k/\mathrm{cm^3\ s^{-1}}) = -12.6$.

(7) τ is defined as the time in which a concentration perturbation falls to $1/\mathrm{e}$ of its initial value.
(8) Note that the term Arrhenius activation energy is to be used only for the empirical quantity defined in the table. Other empirical equations with different 'activation energies', such as $k(T) = A'T^n \exp(-E'_a/RT)$, are also being used.

The term activation energy is also used for an energy threshold appearing in the electronic potential (the height of the electronic energy barrier). For this 'activation energy' the symbol E_0 and the term threshold energy is preferred, but E_a is also commonly used. Furthermore, E_0 may or may not include a correction for zero point energies of reactants and the transition states.

It is thus recommended to specify in any given context exactly which activation energy is meant and to reserve (Arrhenius) activation energy only and exactly for the quantity defined in the table.

Name	*Symbol*	*Definition*	*SI unit*	*Notes*
pre-exponential factor, frequency factor	A	$k = A\exp(-E_a/RT)$	$(m^3\ mol^{-1})^{n-1}\ s^{-1}$	
volume of activation	$\Delta^{\ddagger}V, \Delta V^{\ddagger}$	$\Delta^{\ddagger}V = -RT(\partial \ln k/\partial T)$	$m^3\ mol^{-1}$	
hard sphere radius	r		m	
collision diameter	d	$d_{AB} = r_A + r_B$	m	
collision cross section	σ	$\sigma = \pi d_{AB}{}^2$	m^2	
mean relative speed between A and B	$\bar{c}_{AB}$	$\bar{c}_{AB} = (8kT/\pi\mu)^{1/2}$	$m\ s^{-1}$	9
collision frequency				
of A with A	$z_A(A)$	$z_A(A) = \sqrt{2}C_A\sigma\bar{c}$	s^{-1}	10
of A with B	$z_A(B)$	$z_A(B) = C_B\sigma\bar{c}_{AB}$	s^{-1}	10
collision density, collision number				
of A with A	Z_{AA}	$Z_{AA} = C_A z_A(A)$	$s^{-1}\ m^{-3}$	11
of A with B	Z_{AB}	$Z_{AB} = C_A z_A(B)$	$s^{-1}\ m^{-3}$	11
collision frequency factor	z_{AB}	$z_{AB} = Z_{AB}/Lc_A c_B$	$m^3\ mol^{-1}\ s^{-1}$	11
mean free path	λ	$\lambda = \bar{c}/z_A$	m	
impact parameter	b		m	12
scattering angle	θ		1, rad	13
differential cross section	I_{ji}	$I_{ji} = d\sigma_{ji}/d\Omega$	$m^2\ sr^{-1}$	14
total cross section	σ_{ji}	$\sigma_{ji} = \int I_{ji}\, d\Omega$	m^2	14
scattering matrix	$\boldsymbol{S}$		1	15
transition probability	P_{ji}	$P_{ji} = \lvert S_{ji}\rvert^2$	1	14, 15
standard enthalpy of activation	$\Delta^{\ddagger}H^{\circ}, \Delta H^{\ddagger}$		$J\ mol^{-1}$	16

(9) μ is the reduced mass.
(10) C denotes the number concentration.
(11) Z_{AA} and Z_{AB} are the total number of AA or AB collisions per time and volume in a system containing only A molecules, or containing two types of molecules A and B. Three-body collisions can be treated in a similar way.
(12) The impact parameter b characterizes an individual collision between two particles; it is defined as the distance of closest approach that would result if the particle trajectories were undeflected by the collision.
(13) $\theta = 0$ implies no deflection.
(14) In all these matrix quantities the first index refers to the final and the second to the initial channel. i and j denote reactant and product channels, respectively, and Ω denotes solid angle; $d\sigma_{ji}/d\Omega$ = (scattered particle current per solid angle)/(incident particle current per area). Elastic scattering implies $i = j$. Both I_{ji} and σ_{ji} depend on the total energy of relative motion, and may be written $I_{ji}(E)$ and $\sigma_{ji}(E)$.
(15) The scattering matrix $\boldsymbol{S}$ is used in quantum discussions of scattering theory; S_{ji} is equal to the ratio (total probability current scattered in channel j)/(total probability current incident in channel i). $\boldsymbol{S}$ is a unitary matrix $\boldsymbol{SS}^{\dagger} = \boldsymbol{1}$. P_{ji} is the probability that collision partners incident in channel i will emerge in channel j.
(16) The quantities $\Delta^{\ddagger}H^{\circ}$, $\Delta^{\ddagger}U^{\circ}$, $\Delta^{\ddagger}S^{\circ}$ and $\Delta^{\ddagger}G^{\circ}$ are used in the transition state theory of chemical reaction. They are normally used only in connection with elementary reactions. The relation between the rate constant k and these quantities is

$$k = \kappa(k_B T/h)\exp(-\Delta^{\ddagger}G^{\circ}/RT),$$

where k has the dimensions of a first-order rate constant and is obtained by multiplication of an nth-order rate constant by $(c^{\circ})^{n-1}$. κ is a transmission coefficient, and $\Delta^{\ddagger}G^{\circ} = \Delta^{\ddagger}H^{\circ} - T\Delta^{\ddagger}S^{\circ}$. Unfortunately the standard symbol $^{\circ}$ is usually omitted, and these quantities are usually written $\Delta H^{\ddagger}$, $\Delta U^{\ddagger}$, $\Delta S^{\ddagger}$ and $\Delta G^{\ddagger}$.

Name	*Symbol*	*SI unit*	*Notes*
standard internal energy of activation	$\Delta^{\ddagger}U^{\ominus}, \Delta U^{\ddagger}$	$J\ mol^{-1}$	16
standard entropy of activation	$\Delta^{\ddagger}S^{\ominus}, \Delta S^{\ddagger}$	$J\ mol^{-1}\ K^{-1}$	16
standard Gibbs energy of activation	$\Delta^{\ddagger}G^{\ominus}, \Delta G^{\ddagger}$	$J\ mol^{-1}$	16
quantum yield, photochemical yield	ϕ, Φ	1	17

(17) The quantum yield ϕ is defined in general by [28]

$$\phi = \frac{\text{number of defined events}}{\text{number of photons absorbed}}$$

For a photochemical reaction it can be defined as

$$\phi = \frac{\text{rate of conversion}}{\text{rate of photon absorption}} = \frac{\mathrm{d}\xi/\mathrm{d}t}{\mathrm{d}n_\gamma/\mathrm{d}t}$$

2.13 ELECTROCHEMISTRY

Electrochemical concepts, terminology and symbols are more extensively described in [1.i]. For the field of semiconductor electrochemistry and photoelectrochemical energy conversion see [29] and for corrosion nomenclature [30].

Name	*Symbol*	*Definition*	*SI unit*	*Notes*
elementary charge, (proton charge)	e		C	
Faraday constant	F	$F = eL$	$\mathrm{C\,mol^{-1}}$	
charge number of an ion	z	$z_B = Q_B/e$	1	1
ionic strength,				
molality basis	I_m, I	$I_m = \frac{1}{2}\sum m_i z_i^2$	$\mathrm{mol\,kg^{-1}}$	
concentration basis	I_c, I	$I_c = \frac{1}{2}\sum c_i z_i^2$	$\mathrm{mol\,m^{-3}}$	2
mean ionic activity	$a_\pm$	$a_\pm = m_\pm \gamma_\pm / m^\ominus$	1	3, 4
activity of an electrolyte	$a(\mathrm{A}_{\nu_+}\mathrm{B}_{\nu_-})$	$a(\mathrm{A}_{\nu_+}\mathrm{B}_{\nu_-}) = a_\pm^{(\nu_+ + \nu_-)}$	1	3
mean ionic molality	$m_\pm$	$m_\pm^{(\nu_+ + \nu_-)} = m_+^{\nu_+} m_-^{\nu_-}$	$\mathrm{mol\,kg^{-1}}$	3
mean ionic activity coefficient	$\gamma_\pm$	$\gamma_\pm^{(\nu_+ + \nu_-)} \doteq \gamma_+^{\nu_+} \gamma_-^{\nu_-}$	1	3
charge number of electrochemical cell reaction	n, ν_e, z		1	5
electric potential difference (of a galvanic cell)	$\Delta V, U, E$	$\Delta V = V_R - V_L$	V	6
emf, electromotive force	E	$E = \lim_{I \to 0} \Delta V$	V	7
standard emf, standard potential of the electrochemical cell reaction	$E^\ominus$	$E^\ominus = -\Delta_r G^\ominus / nF = (RT/nF) \ln K^\ominus$	V	4, 8

(1) The definition applies to entities B.
(2) To avoid confusion with the cathodic current, symbol I_c (note roman subscript), the symbol I or sometimes μ (when the current is denoted by I) is used for ionic strength based on concentration.
(3) ν_+ and ν_- are the numbers of cations and anions per formula unit of an electrolyte $\mathrm{A}_{\nu+}\mathrm{B}_{\nu-}$.

Example For $Al_2(SO_4)_3$, $\nu_+ = 2$ and $\nu_- = 3$.

m_+ and m_-, and γ_+ and γ_-, are the separate cation and anion molalities and activity coefficients. If the molality of $\mathrm{A}_{\nu+}\mathrm{B}_{\nu-}$ is m, then $m_+ = \nu_+ m$ and $m_- = \nu_- m$. A similar definition is used on a concentration scale for the mean ionic concentration $c_\pm$.
(4) The symbol $\ominus$ or $\circ$ is used to indicate standard. They are equally acceptable.
(5) n is he number of electrons transferred according to the cell reaction (or half-cell reactions) as written; n is a positive integer.
(6) V_R and V_L are the potentials of the electrodes shown on the right- and left-hand sides, respectively, in the diagram representing the cell. When ΔV is positive, positive charge flows from left to right through the cell, and from right to left in the external circuit, if the cell is short-circuited.
(7) The definition of emf is discussed on p.60. The symbol E_{MF} is no longer recommended for this quantity.
(8) $\Delta_r G^\ominus$ and $K^\ominus$ apply to the cell reaction in the direction in which reduction occurs at the right-hand electrode and oxidation at the left-hand electrode, in the diagram representing the cell (see p.60). (Note the mnemonic 'reduction at the right'.)

Name	*Symbol*	*Definition*	*SI unit*	*Notes*
standard electrode potential	E°		V	4, 9
emf of the cell, potential of the electrochemical cell reaction	E	$E = E^\circ - (RT/nF)\sum \nu_i \ln a_i$	V	10
pH	pH	$\mathrm{pH} \approx -\lg\left[\frac{c(\mathrm{H}^+)}{\mathrm{mol\,dm^{-3}}}\right]$	1	11
inner electric potential	ϕ	$\nabla\phi = -\boldsymbol{E}$	V	12
outer electric potential	ψ	$\psi = Q/4\pi\varepsilon_0 r$	V	13
surface electric potential	χ	$\chi = \phi - \psi$	V	
Galvani potential difference	$\Delta\phi$	$\Delta_\alpha^\beta\phi = \phi^\beta - \phi^\alpha$	V	14
Volta potential difference	$\Delta\psi$	$\Delta_\alpha^\beta\psi = \psi^\beta - \psi^\alpha$	V	15
electrochemical potential	$\tilde{\mu}$	$\tilde{\mu}_\mathrm{B}{}^\alpha = (\partial G/\partial n_\mathrm{B}{}^\alpha)$	$\mathrm{J\,mol^{-1}}$	1, 16
electric current	I	$I = \mathrm{d}Q/\mathrm{d}t$	A	17
(electric) current density	j	$j = I/A$	$\mathrm{A\,m^{-2}}$	17
(surface) charge density	σ	$\sigma = Q/A$	$\mathrm{C\,m^{-2}}$	
electrode reaction rate constant	k	$k_\mathrm{ox} = I_\mathrm{a}/(nFA\prod_i c_i{}^{n_i})$	(varies)	18, 19
mass transfer coefficient, diffusion rate constant	k_d	$k_{\mathrm{d,B}} = \lvert\nu_\mathrm{B}\rvert I_{\mathrm{l,B}}/nFcA$	$\mathrm{m\,s^{-1}}$	1, 19
thickness of diffusion layer	δ	$\delta_\mathrm{B} = D_\mathrm{B}/k_{\mathrm{d,B}}$	m	1

(9) Standard potential of an electrode reaction, abbreviated as standard electrode potential, is the value of the standard emf of a cell in which molecular hydrogen is oxidized to solvated protons at the left-hand electrode. For example, the standard potential of the Zn^{2+}/Zn electrode, denoted $E^\circ(Zn^{2+}/Zn)$, is the emf of the cell in which the reaction $Zn^{2+}(aq) + H_2 \rightarrow 2H^+(aq) + Zn$ takes place under standard conditions (see p.61). The concept of an *absolute* electrode potential is discussed in reference [31].

(10) $\sum \nu_i \ln a_i$ refers to the cell reaction, with ν_i positive for products and negative for reactants; for the complete cell reaction only mean ionic activities $a_\pm$ are involved.

(11) The precise definition of pH is discussed on p.62. The symbol pH is an exception to the general rules for the symbols of physical quantities (p.5) in that it is a two-letter symbol and it is always printed in roman (upright) type.

(12) $\boldsymbol{E}$ is the electric field strength within the phase concerned.

(13) The definition is an example specific to a conducting sphere of excess charge Q and radius r.

(14) $\Delta\phi$ is the electric potential difference between points within the bulk phases α and β; it is measurable only if the phases are of identical composition.

(15) $\Delta\psi$ is the electric potential difference due to the charge on phases α and β. It is measurable or calculable by classical electrostatics from the charge distribution.

(16) The chemical potential is related to the electrochemical potential by the equation $\mu_\mathrm{B}{}^\alpha = \tilde{\mu}_\mathrm{B}{}^\alpha - z_\mathrm{B}F\phi^\alpha$. For an uncharged species, $z_\mathrm{B} = 0$, the electrochemical potential is equal to the chemical potential.

(17) I, j and α may carry one of the subscripts: a for anodic, c for cathodic, e or o for exchange, or l for limiting. I_a and I_c are the anodic and cathodic partial currents. The cathode is the electrode where reduction takes place, and the anode is the electrode where oxidation takes place.

(18) For reduction the rate constant k_red can be defined analogously in terms of the cathodic current I_c. For first-order reaction the SI unit is $\mathrm{m\,s^{-1}}$. n_i is the order of reaction with respect to component i.

(19) For more information on kinetics of electrode reactions and on transport phenomena in electrolyte systems see [32] and [33].

Name	*Symbol*	*Definition*	*SI unit*	*Notes*
transfer coefficient (electrochemical)	α	$\alpha_c = \dfrac{-\lvert\nu\rvert RT}{nF}\dfrac{\partial \ln\lvert I_c\rvert}{\partial E}$	1	17, 19
overpotential,	η	$\eta = E_I - E_{I=0} - IR_u$	V	19
electrokinetic potential, (zeta potential)	ζ		V	
conductivity	$\kappa, (\sigma)$	$\kappa = j/E$	$S\,m^{-1}$	12, 20
conductivity cell constant	K_{cell}	$K_{cell} = \kappa R$	m^{-1}	
molar conductivity (of an electrolyte)	Λ	$\Lambda_B = \kappa/c_B$	$S\,m^2\,mol^{-1}$	1, 21
electric mobility	$u, (\mu)$	$u_B = v_B/E$	$m^2\,V^{-1}\,s^{-1}$	1, 22
ionic conductivity, molar conductivity of an ion	λ	$\lambda_B = \lvert z_B\rvert F u_B$	$S\,m^2\,mol^{-1}$	1, 23
transport number	t	$t_B = j_B/\sum j_i$	1	1
reciprocal radius of ionic atmosphere	κ	$\kappa = (2F^2 I_c/\varepsilon RT)^{1/2}$	m^{-1}	24

(20) Conductivity was formerly called specific conductance.
(21) The unit $S\,cm^2\,mol^{-1}$ is often used for molar conductivity.
(22) v_B is the speed of entities B and E is the electric field strength within the phase concerned.
(23) It is important to specify the entity to which molar conductivity refers; thus for example $\lambda(Mg^{2+}) = 2\lambda(\frac{1}{2}Mg^{2+})$. It is standard practice to choose the entity to be $1/z_B$ of an ion of charge number z_B, so that for example molar conductivities for potassium, barium and lanthanum ions would be quoted as $\lambda(K^+)$, $\lambda(\frac{1}{2}Ba^{2+})$, or $\lambda(\frac{1}{3}La^{3+})$.
(24) κ appears in Debye–Hückel theory. The Debye length, $L_D = \kappa^{-1}$, appears in Gouy–Chapman theory, and in the theory of semiconductor space charge. I_c is the ionic strength.

Conventions concerning the signs of electric potential differences, electromotive forces, and electrode potentials[1]

(i) The electric potential difference for a galvanic cell

The cell should be represented by a diagram, for example:

$$Zn|Zn^{2+}\,\vdots\,Cu^{2+}|Cu$$

A single vertical bar (|) should be used to represent a phase boundary, a dashed vertical bar (¦) to represent a junction between miscible liquids, and double dashed vertical bars (¦¦) to represent a liquid junction in which the liquid junction potential is assumed to be eliminated. The electric potential difference, denoted ΔV or E, is equal in sign and magnitude to the electric potential of a metallic conducting lead on the right minus that of a similar lead on the left. The emf (electromotive force), also usually denoted E, is the limiting value of the electric potential difference for zero current through the cell, all local charge transfer equilibria and chemical equilibria being established. Note that the symbol E is often used for both the potential difference and the emf, and this can sometimes lead to confusion.

(1) These are in accordance with the 'Stockholm Convention' of 1953 [34].

When the reaction of the cell is written as

$$\tfrac{1}{2}Zn + \tfrac{1}{2}Cu^{2+} = \tfrac{1}{2}Zn^{2+} + \tfrac{1}{2}Cu, \qquad n = 1$$

or

$$Zn + Cu^{2+} = Zn^{2+} + Cu, \qquad n = 2,$$

this implies a cell diagram drawn, as above, so that this reaction takes place when positive electricity flows through the cell from left to right (and therefore through the outer part of the circuit from right to left). In the above example the right-hand electrode is positive (unless the ratio $[Cu^{2+}]/[Zn^{2+}]$ is extremely small), so that this is the direction of spontaneous flow if a wire is connected across the two electrodes. If, however, the reaction is written as

$$\tfrac{1}{2}Cu + \tfrac{1}{2}Zn^{2+} = \tfrac{1}{2}Cu^{2+} + \tfrac{1}{2}Zn, \qquad n = 1$$

or

$$Cu + Zn^{2+} = Cu^{2+} + Zn, \qquad n = 2,$$

this implies the cell diagram

$$Cu|Cu^{2+} \vdots Zn^{2+}|Zn$$

and the electric potential difference of the cell so specified will be negative. Thus a cell diagram may be drawn either way round, and correspondingly the electric potential difference appropriate to the diagram may be either positive or negative.

(ii) Electrode potential (potential of an electrode reaction)

The so-called electrode potential of an electrode is defined as the emf of a cell in which the electrode on the left is a standard hydrogen electrode and the electrode on the right is the electrode in question. For example, for the silver/silver chloride electrode (written $Cl^-(aq)|AgCl|Ag$) the cell in question is

$$Pt|H_2(g, p = p^{\ominus})|HCl(aq, a_\pm = 1) \vdots HCl(aq, a_\pm{}')|AgCl|Ag$$

A liquid junction will be necessary in this cell whenever $a_\pm{}'(HCl)$ on the right differs from $a_\pm(HCl)$ on the left. The reaction taking place at the silver/silver chloride electrode is

$$AgCl(s) + e^- \rightarrow Ag(s) + Cl^-(aq)$$

The complete cell reaction is

$$AgCl(s) + \tfrac{1}{2}H_2(g) \rightarrow H^+(aq) + Cl^-(aq) + Ag(s)$$

In the standard state of the hydrogen electrode, $p(H_2) = p^{\ominus} = 10^5$ Pa and $a_\pm(HCl) = 1$, the emf of this cell is the electrode potential of the silver/silver chloride electrode. If, in addition, the mean activity of the HCl in the silver/silver chloride electrode $a_\pm(HCl) = 1$, then the emf is equal to $E^{\ominus}$ for this electrode. The standard electrode potential for $HCl(aq)|AgCl|Ag$ has the value $E^{\ominus} = +0.222\,17$ V at 298.15 K. For $p^{\ominus} = 101\,325$ Pa the standard potential of this electrode (and of any electrode involving only condensed phases) is higher by 0.17 mV; i.e.

$$E^{\ominus}(101\,325 \text{ Pa}) = E^{\ominus}(10^5 \text{ Pa}) + 0.17 \text{ mV}$$

A compilation of standard electrode potentials, and their conversion between different standard pressures, can be found in [29]. Notice that in writing the cell whose emf represents an electrode potential, it is important that the hydrogen electrode should always be on the left.

(iii) Operational definition of pH [36]

The notional definition of pH given in the table above is in practice replaced by the following operational definition. For a solution X the emf $E(\text{X})$ of the galvanic cell

$$\begin{array}{c|c} \text{reference} & \text{KCl(aq,} \\ \text{electrode} & m > 3.5\ \text{mol kg}^{-1}) \end{array} \,\vdots\vdots\, \begin{array}{c|c|c} \text{solution X} & \text{H}_2(\text{g}) & \text{Pt} \end{array}$$

is measured, and likewise the emf $E(\text{S})$ of the cell that differs only by the replacement of the solution X of unknown pH(X) by the solution S of standard pH(S). The unknown pH is then given by

$$\text{pH(X)} = \text{pH(S)} + (E_\text{S} - E_\text{X})F/(RT \ln 10)$$

Thus defined, pH is dimensionless. Values of pH(S) for several standard solutions and temperatures are listed in [36]. The reference value pH standard is an aqueous solution of potassium hydrogen phthalate at a molality of exactly 0.05 mol kg^{-1}: at 25 °C (298.15 K) this has a pH of 4.005.

In practice a glass electrode is almost always used in place of the Pt|H_2 electrode. The cell might then take the form

$$\begin{array}{c|c} \text{reference} & \text{KCl(aq,} \\ \text{electrode} & m > 3.5\ \text{mol kg}^{-1}) \end{array} \,\vdots\vdots\, \begin{array}{c|c|c|c|c} \text{solution X} & \text{glass} & \text{H}^+, \text{Cl}^- & \text{AgCl} & \text{Ag} \end{array}$$

The solution to the right of the glass electrode is usually a buffer solution of KH_2PO_4 and Na_2HPO_4, with 0.1 mol dm^{-3} of NaCl. The reference electrode is usually a calomel electrode, silver/silver chloride electrode, or a thallium amalgam/thallous chloride electrode. The emf of this cell depends on $a(\text{H}^+)$ in the solution X in the same way as that of the cell with the Pt|H_2 electrode, and thus the same procedure is followed.

In the restricted range of dilute aqueous solutions having amount concentrations less than 0.1 mol dm^{-3} and being neither strongly acidic nor strongly alkaline ($2 < \text{pH} < 12$) the above definition is such that

$$\begin{aligned} \text{pH} &= -\text{lg}[\gamma_\pm c(\text{H}^+)/(\text{mol dm}^{-3})] \pm 0.02, \\ &= -\text{lg}[\gamma_\pm m(\text{H}^+)/(\text{mol kg}^{-1})] \pm 0.02, \end{aligned}$$

where $c(\text{H}^+)$ denotes the amount concentration of hydrogen ion H^+ and $m(\text{H}^+)$ the corresponding molality, and $\gamma_\pm$ denotes the mean ionic activity coefficient of a typical uni-univalent electrolyte in the solution on a concentration basis or a molality basis as appropriate. For further information on the definition of pH see [36].

2.14 COLLOID AND SURFACE CHEMISTRY

The recommendations given here are based on more extensive IUPAC recommendations [1.e–h] and [37–39]. Catalyst characterization is described in [40] and quantities related to macromolecules in [41].

Name	*Symbol*	*Definition*	*SI unit*	*Notes*
specific surface area	a, a_s, s	$a = A/m$	$m^2\,kg^{-1}$	
surface amount of B, adsorbed amount of B	$n_B{}^s$, $n_B{}^a$		mol	1
surface excess of B	$n_B{}^\sigma$		mol	2
surface excess concentration of B	Γ_B, $(\Gamma_B{}^\sigma)$	$\Gamma_B = n_B{}^\sigma/A$	$mol\,m^{-2}$	2
total surface excess concentration	Γ, (Γ^σ)	$\Gamma = \sum_i \Gamma_i$	$mol\,m^{-2}$	
area per molecule	a, σ	$a_B = A/N_B{}^\sigma$	m^2	3
area per molecule in a filled monolayer	a_m, σ_m	$a_{m,B} = A/N_{m,B}$	m^2	3
surface coverage	θ	$\theta = N_B{}^\sigma/N_{m,B}$	1	3
contact angle	θ		1, rad	
film thickness	t, h, δ		m	
thickness of (surface or interfacial) layer	τ, δ, t		m	
surface tension, interfacial tension	γ, σ	$\gamma = (\partial G/\partial A_s)_{T,p}$	$N\,m^{-1}$, $J\,m^{-2}$	
film tension	Σ_f	$\Sigma_f = 2\gamma_f$	$N\,m^{-1}$	4
reciprocal thickness of the double layer	κ	$\kappa = (2F^2 I_c/\varepsilon RT)^{\frac{1}{2}}$	m^{-1}	
average molar masses				
number-average	M_n	$M_n = \Sigma n_i M_i/\Sigma n_i$	$kg\,mol^{-1}$	
mass-average	M_m	$M_m = \Sigma n_i M_i{}^2/\Sigma n_i M_i$	$kg\,mol^{-1}$	
Z-average	M_z	$M_z = \Sigma n_i M_i{}^3/\Sigma n_i M_i{}^2$	$kg\,mol^{-1}$	
sedimentation coefficient	s	$s = v/a$	s	5
van der Waals constant	λ		J	
retarded van der Waals constant	β, B		J	
van der Waals–Hamaker constant	A_H		J	
surface pressure	π^s, π	$\pi^s = \gamma^0 - \gamma$	$N\,m^{-1}$	6

(1) The value of $n_B{}^s$ depends on the thickness assigned to the surface layer.
(2) The values of $n_B{}^\sigma$ and Γ_B depend on the convention used to define the position of the Gibbs surface. They are given by the excess amount of B or surface concentration of B over values that would apply if each of the two bulk phases were homogeneous right up to the Gibbs surface. See [1.e], and also additional recommendations on p.64.
(3) $N_B{}^\sigma$ is the number of adsorbed molecules ($N_B{}^\sigma = L n_B{}^\sigma$), and $N_{m,B}$ is the number of adsorbed molecules in a filled monolayer. The definition applies to entities B.
(4) The definition applies only to a symmetrical film, for which the two bulk phases on either side of the film are the same, and γ_f is the surface tension of a film/bulk interface.
(5) In the definition, v is the velocity of sedimentation and a is the acceleration of free fall or centrifugation. The symbol for a limiting sedimentation coefficient is $[s]$, for a reduced sedimentation coefficient s°, and for a reduced limiting sedimentation coefficient $[s^\circ]$; see [1.e] for further details.
(6) In the definition, γ^0 is the surface tension of the clean surface and γ that of the covered surface.

Additional recommendations

The superscript s denotes the properties of a surface or interfacial layer. In the presence of adsorption it may be replaced by the superscript a.

Examples	Helmholtz energy of interfacial layer	A^s
	amount of adsorbed substance	n^a, n^s
	amount of adsorbed O_2	$n^a(O_2)$, $n^s(O_2)$, or $n(O_2, a)$

The subscript m denotes the properties of a monolayer.

Example	area per molecule B in a monolayer	$a_m(B)$

The superscript σ is used to denote a surface excess property relative to the Gibbs surface.

Example	surface excess amount (or Gibbs surface excess of B)	n_B^σ

In general the values of Γ_A and Γ_B depend on the position chosen for the Gibbs dividing surface. However, two quantities, $\Gamma_B^{(A)}$ and $\Gamma_B^{(n)}$ (and correspondingly $n_B^{\sigma\,(A)}$ and $n_B^{\sigma\,(n)}$), may be defined in a way that is invariant to this choice (see [1.e]). $\Gamma_B^{(A)}$ is called the *relative* surface excess concentration of B with respect to A, or more simply the relative adsorption of B; it is the value of Γ_B when the surface is chosen to make $\Gamma_A = 0$. $\Gamma_B^{(n)}$ is called the *reduced* surface excess concentration of B, or more simply the reduced adsorption of B; it is the value of Γ_B when the surface is chosen to make the total excess $\Gamma = \sum_i \Gamma_i = 0$.

Properties of phases (α, β, γ) may be denoted by corresponding superscript indices.

Examples	surface tension of phase α	γ^α
	interfacial tension between phases α and β	$\gamma^{\alpha\beta}$

Symbols of thermodynamic quantities divided by surface area are usually the corresponding lower case letters; an alternative is to use a circumflex.

Example	interfacial entropy per area	$s^s\ (=\hat{s}^s) = S^s/A$

The following abbreviations are used in colloid chemistry:

c.c.c.	critical coagulation concentration
c.m.c.	critical micellization concentration
i.e.p.	isoelectric point
p.z.c.	point of zero charge

2.15 TRANSPORT PROPERTIES

The names and symbols recommended here are in agreement with those recommended by IUPAP [4] and ISO [5.n]. Further information on transport phenomena in electrochemical systems can also be found in [32].

Name	*Symbol*	*Definition*	*SI unit*	*Notes*
flux (of a quantity X)	J_X, J	$J_X = A^{-1}\mathrm{d}X/\mathrm{d}t$	(varies)	1
volume flow rate	q_V, $\dot{V}$	$q_V = \mathrm{d}V/\mathrm{d}t$	$\mathrm{m^3\,s^{-1}}$	
mass flow rate	q_m, $\dot{m}$	$q_m = \mathrm{d}m/\mathrm{d}t$	$\mathrm{kg\,s^{-1}}$	
mass transfer coefficient	k_d		$\mathrm{m\,s^{-1}}$	
heat flow rate	Φ	$\Phi = \mathrm{d}q/\mathrm{d}t$	W	
heat flux	J_q	$J_q = \Phi/A$	$\mathrm{W\,m^{-2}}$	
thermal conductance	G	$G = \Phi/\Delta T$	$\mathrm{W\,K^{-1}}$	
thermal resistance	R	$R = 1/G$	$\mathrm{K\,W^{-1}}$	
thermal conductivity	λ, k	$\lambda = J_q/(\mathrm{d}T/\mathrm{d}l)$	$\mathrm{W\,m^{-1}\,K^{-1}}$	
coefficient of heat transfer	h, (k, K, α)	$h = J_q/\Delta T$	$\mathrm{W\,m^{-2}\,K^{-1}}$	
thermal diffusivity	a	$a = \lambda/\rho c_p$	$\mathrm{m^2\,s^{-1}}$	
diffusion coefficient	D	$D = -J_n/(\mathrm{d}c/\mathrm{d}l)$	$\mathrm{m^2\,s^{-1}}$	

The following symbols are used in the definitions of the dimensionless quantities: mass (m), time (t), volume (V), area (A), density (ρ), speed (v), length (l), viscosity (η), pressure (p), acceleration of free fall (g), cubic expansion coefficient (α), temperature (T), surface tension (γ), speed of sound (c), mean free path (λ), frequency (f), thermal diffusivity (a), coefficient of heat transfer (h), thermal conductivity (k), specific heat capacity at constant pressure (c_p), diffusion coefficient (D), mole fraction (x), mass transfer coefficient (k_d), permeability (μ), electric conductivity (κ), and magnetic flux density (B).

Name	*Symbol*	*Definition*	*SI unit*
Reynolds number	Re	$Re = \rho v l/\eta$	1
Euler number	Eu	$Eu = \Delta p/\rho v^2$	1
Froude number	Fr	$Fr = v/(lg)^{1/2}$	1
Grashof number	Gr	$Gr = l^3 g\alpha\Delta T\rho^2/\eta^2$	1
Weber number	We	$We = \rho v^2 l/\gamma$	1
Mach number	Ma	$Ma = v/c$	1
Knudsen number	Kn	$Kn = \lambda/l$	1
Strouhal number	Sr	$Sr = lf/v$	1
Fourier number	Fo	$Fo = at/l^2$	1
Péclet number	Pe	$Pe = vl/a$	1
Rayleigh number	Ra	$Ra = l^3 g\alpha\Delta T\rho/\eta a$	1
Nusselt number	Nu	$Nu = hl/k$	1

(1) The flux of molecules to a surface, J_N, determines either the rate at which it would be covered if each molecule stuck, or the rate of effusion through a hole in the surface. In studying the exposure, $\int J_N \mathrm{d}t$, of a surface to a gas, surface scientists find it useful to use the product of pressure and time as a measure of the exposure since this product is proportional to the number flux, J_N, times the time $J_N t = (1/4)C\bar{u}t = (\bar{u}/4kT)pt$, where C is the number density of molecules, $\bar{u}$ their average speed, k the Boltzmann constant and T the thermodynamic temperature. The unit langmuir (symbol: L) corresponds to the exposure of a surface to a gas at 10^{-6} torr for 1 second.

Name	*Symbol*	*Definition*	*SI unit*	*Notes*
Stanton number	*St*	$St = h/\rho v c_p$	1	
Fourier number for mass transfer	Fo^*	$Fo^* = Dt/l^2$	1	2
Péclet number for mass transfer	Pe^*	$Pe^* = vl/D$	1	2
Grashof number for mass transfer	Gr^*	$Gr^* = l^3 g \left(\dfrac{\partial \rho}{\partial x}\right)_{T,p} \left(\dfrac{\Delta x \rho}{\eta}\right)$	1	2
Nusselt number for mass transfer	Nu^*	$Nu^* = k_d l/D$	1	2, 3
Stanton number for mass transfer	St^*	$St^* = k_d/v$	1	2
Prandtl number	*Pr*	$Pr = \eta/\rho a$	1	
Schmidt number	*Sc*	$Sc = \eta/\rho D$	1	
Lewis number	*Le*	$Le = a/D$	1	
magnetic Reynolds number	Rm, Re_m	$Rm = v\mu\kappa l$	1	
Alfvén number	*Al*	$Al = v(\rho\mu)^{\frac{1}{2}}/B$	1	
Hartmann number	*Ha*	$Ha = Bl(\kappa/\eta)^{\frac{1}{2}}$	1	
Cowling number	*Co*	$Co = B^2/\mu\rho v^2$	1	

(2) This quantity applies to the transport of matter in binary mixtures.
(3) The name Sherwood number and symbol *Sh* have been widely used for this quantity.

3
Definitions and symbols for units

3.1 THE INTERNATIONAL SYSTEM OF UNITS (SI)

The International System of units (SI) was adopted by the 11th General Conference on Weights and Measures (CGPM) in 1960 [3]. It is a coherent system of units built from seven *SI base units*, one for each of the seven dimensionally independent base quantities (see section 1.2): they are the metre, kilogram, second, ampere, kelvin, mole, and candela, for the dimensions length, mass, time, electric current, thermodynamic temperature, amount of substance, and luminous intensity, respectively. The definitions of the SI base units are given in section 3.2. The *SI derived units* are expressed as products of powers of the base units, analogous to the corresponding relations between physical quantities but with numerical factors equal to unity [3].

In the International System there is only one SI unit for each physical quantity. This is either the appropriate SI base unit itself (see table 3.3) or the appropriate SI derived unit (see tables 3.4 and 3.5). However, any of the approved decimal prefixes, called *SI prefixes*, may be used to construct decimal multiples or submultiples of SI units (see table 3.6).

It is recommended that only SI units be used in science and technology (with SI prefixes where appropriate). Where there are special reasons for making an exception to this rule, it is recommended always to define the units used in terms of SI units.

3.2 DEFINITIONS OF THE SI BASE UNITS [3]

metre: The metre is the length of path travelled by light in vacuum during a time interval of 1/299 792 458 of a second (17th CGPM, 1983).

kilogram: The kilogram is the unit of mass; it is equal to the mass of the international prototype of the kilogram (3rd CGPM, 1901).

second: The second is the duration of 9 192 631 770 periods of the radiation corresponding to the transition between the two hyperfine levels of the ground state of the caesium-133 atom (13th CGPM, 1967).

ampere: The ampere is that constant current which, if maintained in two straight parallel conductors of infinite length, of negligible circular cross-section, and placed 1 metre apart in vacuum, would produce between these conductors a force equal to 2×10^{-7} newton per metre of length (9th CGPM, 1948).

kelvin: The kelvin, unit of thermodynamic temperature, is the fraction 1/273.16 of the thermodynamic temperature of the triple point of water (13th CGPM, 1967).

mole: The mole is the amount of substance of a system which contains as many elementary entities as there are atoms in 0.012 kilogram of carbon-12. When the mole is used, the elementary entities must be specified and may be atoms, molecules, ions, electrons, other particles, or specified groups of such particles (14th CGPM, 1971).

Examples of the use of the mole

1 mol of H_2 contains about 6.022×10^{23} H_2 molecules, or 12.044×10^{23} H atoms
1 mol of HgCl has a mass of 236.04 g
1 mol of Hg_2Cl_2 has a mass of 472.08 g
1 mol of $Hg_2{}^{2+}$ has a mass of 401.18 g and a charge of 192.97 kC
1 mol of $Fe_{0.91}S$ has a mass of 82.88 g
1 mol of e^- has a mass of 548.60 μg and a charge of −96.49 kC
1 mol of photons whose frequency is 5×10^{14} Hz has energy of about 199.5 kJ

See also section 2.10, p.46.

candela: The candela is the luminous intensity, in a given direction, of a source that emits monochromatic radiation of frequency 540×10^{12} hertz and that has a radiant intensity in that direction of (1/683) watt per steradian (16th CGPM, 1979).

3.3 NAMES AND SYMBOLS FOR THE SI BASE UNITS

The symbols listed here are internationally agreed and should not be changed in other languages or scripts. See sections 1.3 and 1.4 on the printing of symbols for units. Recommended representations for these symbols for use in systems with limited character sets can be found in [7].

Physical quantity	*Name of SI unit*	*Symbol for SI unit*
length	metre	m
mass	kilogram	kg
time	second	s
electric current	ampere	A
thermodynamic temperature	kelvin	K
amount of substance	mole	mol
luminous intensity	candela	cd

3.4 SI DERIVED UNITS WITH SPECIAL NAMES AND SYMBOLS

Physical quantity	*Name of SI unit*	*Symbol for SI unit*	*Expression in terms of SI base units*	
frequency[1]	hertz	Hz	s^{-1}	
force	newton	N	$m\ kg\ s^{-2}$	
pressure, stress	pascal	Pa	$N\ m^{-2}$	$= m^{-1}\ kg\ s^{-2}$
energy, work, heat	joule	J	N m	$= m^2\ kg\ s^{-2}$
power, radiant flux	watt	W	$J\ s^{-1}$	$= m^2\ kg\ s^{-3}$
electric charge	coulomb	C	A s	
electric potential, electromotive force	volt	V	$J\ C^{-1}$	$= m^2\ kg\ s^{-3}\ A^{-1}$
electric resistance	ohm	Ω	$V\ A^{-1}$	$= m^2\ kg\ s^{-3}\ A^{-2}$
electric conductance	siemens	S	Ω^{-1}	$= m^{-2}\ kg^{-1}\ s^3\ A^2$
electric capacitance	farad	F	$C\ V^{-1}$	$= m^{-2}\ kg^{-1}\ s^4\ A^2$
magnetic flux density	tesla	T	$V\ s\ m^{-2}$	$= kg\ s^{-2}\ A^{-1}$
magnetic flux	weber	Wb	V s	$= m^2\ kg\ s^{-2}\ A^{-1}$
inductance	henry	H	$V\ A^{-1}\ s$	$= m^2\ kg\ s^{-2}\ A^{-2}$
Celsius temperature[2]	degree Celsius	°C	K	
luminous flux	lumen	lm	cd sr	
illuminance	lux	lx	$cd\ sr\ m^{-2}$	
activity[3] (radioactive)	becquerel	Bq	s^{-1}	
absorbed dose[3] (of radiation)	gray	Gy	$J\ kg^{-1}$	$= m^2\ s^{-2}$
dose equivalent[3] (dose equivalent index)	sievert	Sv	$J\ kg^{-1}$	$= m^2\ s^{-2}$
plane angle[4]	radian	rad	1	$= m\ m^{-1}$
solid angle[4]	steradian	sr	1	$= m^2\ m^{-2}$

(1) For radial (angular) frequency and for angular velocity the unit rad s^{-1}, or simply s^{-1}, should be used, and this may *not* be simplified to Hz. The unit Hz should be used *only* for frequency in the sense of cycles per second.
(2) The Celsius temperature θ is defined by the equation

$$\theta/°C = T/K - 273.15$$

The SI unit of Celsius temperature is the degree Celsius, °C, which is equal to the kelvin, K. °C should be treated as a single symbol, with no space between the ° sign and the letter C. (The symbol °K, and the symbol °, should no longer be used.)
(3) The units becquerel, gray and sievert are admitted for reasons of safeguarding human health [3].
(4) The units radian and steradian are described as 'SI supplementary units' [3]. However, in chemistry, as well as in physics [4], they are usually treated as dimensionless derived units, and this was recognized by CIPM in 1980. Since they are then of dimension 1, this leaves open the possibility of including them or omitting them in expressions of SI derived units. In practice this means that rad and sr may be used when appropriate and may be omitted if clarity is not lost thereby.

3.5 SI DERIVED UNITS FOR OTHER QUANTITIES

This table gives examples of other SI derived units; the list is merely illustrative.

Physical quantity	*Expression in terms of SI base units*	
area	m^2	
volume	m^3	
speed, velocity	$m\,s^{-1}$	
angular velocity	s^{-1}, $rad\,s^{-1}$	
acceleration	$m\,s^{-2}$	
moment of force	$N\,m$	$= m^2\,kg\,s^{-2}$
wavenumber	m^{-1}	
density, mass density	$kg\,m^{-3}$	
specific volume	$m^3\,kg^{-1}$	
amount concentration[1]	$mol\,m^{-3}$	
molar volume	$m^3\,mol^{-1}$	
heat capacity, entropy	$J\,K^{-1}$	$= m^2\,kg\,s^{-2}\,K^{-1}$
molar heat capacity, molar entropy	$J\,K^{-1}\,mol^{-1}$	$= m^2\,kg\,s^{-2}\,K^{-1}\,mol^{-1}$
specific heat capacity, specific entropy	$J\,K^{-1}\,kg^{-1}$	$= m^2\,s^{-2}\,K^{-1}$
molar energy	$J\,mol^{-1}$	$= m^2\,kg\,s^{-2}\,mol^{-1}$
specific energy	$J\,kg^{-1}$	$= m^2\,s^{-2}$
energy density	$J\,m^{-3}$	$= m^{-1}\,kg\,s^{-2}$
surface tension	$N\,m^{-1} = J\,m^{-2}$	$= kg\,s^{-2}$
heat flux density, irradiance	$W\,m^{-2}$	$= kg\,s^{-3}$
thermal conductivity	$W\,m^{-1}\,K^{-1}$	$= m\,kg\,s^{-3}\,K^{-1}$
kinematic viscosity, diffusion coefficient	$m^2\,s^{-1}$	
dynamic viscosity	$N\,s\,m^{-2} = Pa\,s$	$= m^{-1}\,kg\,s^{-1}$
electric charge density	$C\,m^{-3}$	$= m^{-3}\,s\,A$
electric current density	$A\,m^{-2}$	
conductivity	$S\,m^{-1}$	$= m^{-3}\,kg^{-1}\,s^3\,A^2$
molar conductivity	$S\,m^2\,mol^{-1}$	$= kg^{-1}\,mol^{-1}\,s^3\,A^2$
permittivity	$F\,m^{-1}$	$= m^{-3}\,kg^{-1}\,s^4\,A^2$
permeability	$H\,m^{-1}$	$= m\,kg\,s^{-2}\,A^{-2}$
electric field strength	$V\,m^{-1}$	$= m\,kg\,s^{-3}\,A^{-1}$
magnetic field strength	$A\,m^{-1}$	
luminance	$cd\,m^{-2}$	
exposure (X and γ rays)	$C\,kg^{-1}$	$= kg^{-1}\,s\,A$
absorbed dose rate	$Gy\,s^{-1}$	$= m^2\,s^{-3}$

(1) The words 'amount concentration' are an abbreviation for 'amount-of-substance concentration'. When there is not likely to be any ambiguity this quantity may be called simply 'concentration'.

3.6 SI PREFIXES

To signify decimal multiples and submultiples of SI units the following prefixes may be used [3].

Submultiple	*Prefix*	*Symbol*	*Multiple*	*Prefix*	*Symbol*
10^{-1}	deci	d	10	deca	da
10^{-2}	centi	c	10^{2}	hecto	h
10^{-3}	milli	m	10^{3}	kilo	k
10^{-6}	micro	μ	10^{6}	mega	M
10^{-9}	nano	n	10^{9}	giga	G
10^{-12}	pico	p	10^{12}	tera	T
10^{-15}	femto	f	10^{15}	peta	P
10^{-18}	atto	a	10^{18}	exa	E
10^{-21}	zepto	z	10^{21}	zetta	Z
10^{-24}	yocto	y	10^{24}	yotta	Y

Prefix symbols should be printed in roman (upright) type with no space between the prefix and the unit symbol.

Example kilometre, km

When a prefix is used with a unit symbol, the combination is taken as a new symbol that can be raised to any power without the use of parentheses.

Examples $1\ \mathrm{cm}^3 = (0.01\ \mathrm{m})^3 = 10^{-6}\ \mathrm{m}^3$
$1\ \mu\mathrm{s}^{-1} = (10^{-6}\ \mathrm{s})^{-1} = 10^{6}\ \mathrm{s}^{-1}$
$1\ \mathrm{V/cm} = 100\ \mathrm{V/m}$
$1\ \mathrm{mmol/dm}^3 = 1\ \mathrm{mol\,m}^{-3}$

A prefix should never be used on its own, and prefixes are not to be combined into compound prefixes.

Example pm, not μμm

The names and symbols of decimal multiples and submultiples of the SI base unit of mass, the kg, which already contains a prefix, are constructed by adding the appropriate prefix to the word gram and symbol g.

Examples mg, not μkg; Mg, not kkg

The SI prefixes are not to be used with °C.

ISO has recommended standard representations of the prefix symbols for use with limited character sets [7].

3.7 UNITS IN USE TOGETHER WITH THE SI

These units are not part of the SI, but it is recognized that they will continue to be used in appropriate contexts. SI prefixes may be attached to some of these units, such as millilitre, ml; millibar, mbar; megaelectronvolt, MeV; kilotonne, kt. A more extensive list of non-SI units, with conversion factors to the corresponding SI units, is given in chapter 7.

Physical quantity	*Name of unit*	*Symbol for unit*	*Value in SI units*
time	minute	min	60 s
time	hour	h	3600 s
time	day	d	86 400 s
plane angle	degree	°	$(\pi/180)$ rad
plane angle	minute	′	$(\pi/10\,800)$ rad
plane angle	second	″	$(\pi/648\,000)$ rad
length	ångström[1]	Å	10^{-10} m
area	barn	b	10^{-28} m^2
volume	litre	l, L	$dm^3 = 10^{-3}$ m^3
mass	tonne	t	$Mg = 10^3$ kg
pressure	bar[1]	bar	10^5 Pa $= 10^5$ N m^{-2}
energy	electronvolt[2]	eV $(= e \times V)$	$\approx 1.60218 \times 10^{-19}$ J
mass	unified atomic mass unit[2,3]	u $(= m_a(^{12}C)/12)$	$\approx 1.66054 \times 10^{-27}$ kg

(1) The ångström and the bar are approved by CIPM [3] for 'temporary use with SI units', until CIPM makes a further recommendation. However, they should not be introduced where they are not used at present.
(2) The values of these units in terms of the corresponding SI units are not exact, since they depend on the values of the physical constants e (for the electronvolt) and N_A (for the unified atomic mass unit), which are determined by experiment. See chapter 5.
(3) The unified atomic mass unit is also sometimes called the dalton, with symbol Da, although the name and symbol have not been approved by CGPM.

3.8 ATOMIC UNITS [9] (see also section 7.3, p.120)

For the purposes of quantum mechanical calculations of electronic wavefunctions, it is convenient to regard certain fundamental constants (and combinations of such constants) as though they were units. They are customarily called *atomic units* (abbreviated: au), and they may be regarded as forming a coherent system of units for the calculation of electronic properties in theoretical chemistry, although there is no authority from CGPM for treating them as units. They are discussed further in relation to the electromagnetic units in chapter 7, p.120. The first five atomic units in the table below have special names and symbols. Only four of these are independent; all others may be derived by multiplication and division in the usual way, and the table includes a number of examples.

The relation of atomic units to the corresponding SI units involves the values of the fundamental physical constants, and is therefore not exact. The numerical values in the table are based on the estimates of the fundamental constants given in chapter 5. The numerical results of calculations in theoretical chemistry are frequently quoted in atomic units, or as numerical values in the form (*physical quantity*)/(*atomic unit*), so that the reader may make the conversion using the current best estimates of the physical constants.

Physical quantity	*Name of unit*	*Symbol for unit*	*Value of unit in SI*
mass	electron rest mass	m_e	$9.109\,3897(54) \times 10^{-31}$ kg
charge	elementary charge	e	$1.602\,177\,33(49) \times 10^{-19}$ C
action	Planck constant/2π[1]	$\hbar$	$1.054\,572\,66(63) \times 10^{-34}$ J s
length	bohr[1]	a_0	$5.291\,772\,49(24) \times 10^{-11}$ m
energy	hartree[1]	E_h	$4.359\,7482(26) \times 10^{-18}$ J
time		$\hbar/E_h$	$2.418\,884\,3341(29) \times 10^{-17}$ s
velocity[2]		$a_0E_h/\hbar$	$2.187\,691\,42(10) \times 10^{6}$ m s^{-1}
force		E_h/a_0	$8.238\,7295(25) \times 10^{-8}$ N
momentum, linear		$\hbar/a_0$	$1.992\,8534(12) \times 10^{-24}$ N s
electric current		$eE_h/\hbar$	$6.623\,6211(20) \times 10^{-3}$ A
electric field		E_h/ea_0	$5.142\,2082(15) \times 10^{11}$ V m^{-1}
electric dipole moment		ea_0	$8.478\,3579(26) \times 10^{-30}$ C m
magnetic flux density		$\hbar/ea_0^2$	$2.350\,518\,08(71) \times 10^{5}$ T
magnetic dipole moment[3]		$e\hbar/m_e$	$1.854\,803\,08(62) \times 10^{-23}$ J T^{-1}

(1) $\hbar = h/2\pi$; $a_0 = 4\pi\varepsilon_0\hbar^2/m_ee^2$; $E_h = \hbar^2/m_ea_0^2$.
(2) The numerical value of the speed of light, when expressed in atomic units, is equal to the reciprocal of the fine structure constant α; c/(au of velocity) $= c\hbar/a_0E_h = \alpha^{-1} \approx 137.035\,9895(61)$.
(3) The atomic unit of magnetic dipole moment is twice the Bohr magneton, μ_B.

3.9 DIMENSIONLESS QUANTITIES

Values of dimensionless physical quantities, more properly called 'quantities of dimension one', are often expressed in terms of mathematically exactly defined values denoted by special symbols or abbreviations, such as % (percent) and ppm (part per million). These symbols are then treated as units, and are used as such in calculations.

Fractions (relative values, yields, efficiencies)

Fractions such as relative uncertainty, mole fraction x (also called amount fraction, or number fraction), mass fraction w, and volume fraction ϕ (see p.41 for all these quantities), are sometimes expressed in terms of the symbols summarized in the table below.

Name	*Symbol*	*Value*	*Examples*
percent	%	10^{-2}	The isotopic abundance of carbon-13 expressed as a mole fraction is $x = 1.1\%$
part per million	ppm	10^{-6}	The relative uncertainty in the Planck constant h ($= 6.626\,0755(40) \times 10^{-34}$ J s) is 0.60 ppm
			The mass fraction of impurities in a sample of copper was found to be less than 3 ppm, $w < 3$ ppm

These multiples of the unit one are not part of the SI and ISO recommends that these symbols should never be used. They are also frequently used as units of 'concentration' without a clear indication of the type of fraction implied (e.g. mole fraction, mass fraction or volume fraction). To avoid ambiguity they should only be used in a context where the meaning of the quantity is carefully defined. Even then, the use of an appropriate SI unit ratio may be preferred.

Further examples: (i) The mass fraction $w = 1.5 \times 10^{-6} = 1.5$ ppm, or $w = 1.5$ mg/kg
(ii) The mole fraction $x = 3.7 \times 10^{-2} = 3.7\%$ or $x = 37$ mmol/mol
(iii) Atomic absorption spectroscopy shows the aqueous solution to contain a mass concentration of nickel $\rho(\text{Ni}) = 2.6\ \text{mg dm}^{-3}$, which is approximately equivalent to a mass fraction $w(\text{Ni}) = 2.6 \times 10^{-6} = 2.6$ ppm.

Note the importance of using the recommended name and symbol for the quantity in each of the above examples. Statements such as 'the concentration of nickel was 2.6 ppm' are ambiguous and should be avoided.

Example (iii) illustrates the approximate equivalence of (ρ/mg dm^{-3}) and (w/ppm) in aqueous solution, which follows from the fact that the mass density of a dilute aqueous solution is always approximately 1.0 g cm^{-3}. Dilute solutions are often measured or calibrated to a known mass concentration in mg dm^{-3}, and this unit is then to be preferred to using ppm to specify a mass fraction.

Deprecated usage

Adding extra labels to ppm and similar symbols, such as ppmv (meaning ppm by volume) should be avoided. Qualifying labels may be added to symbols for physical quantities, but never to units.

Examples: A volume fraction $\phi = 2$ ppm, but *not* a concentration of 2 ppmv.
A mass fraction $w = 0.5\%$, but *not* 0.5% w/w.

The symbols % and ppm should not be used in combination with other units. In table headings and in labelling the axes of graphs the use of % and ppm in the denominator is to be avoided. Although one would write $x(^{13}C) = 1.1\%$, the notation $100\,x$ is to be preferred to $x/\%$ in tables and graphs (see for example table 6.3 on p.98).

The further symbols listed in the table below are also to be found in the literature, but their use is to be deprecated. Note that the names and symbols for 10^{-9} and 10^{-12} in this table are based on the American system of names. In other parts of the world a billion sometimes stands for 10^{12} and a trillion for 10^{18}. Note also that the symbol ppt is sometimes used for part per thousand, and sometimes for part per trillion.

To avoid ambiguity the symbols ppb, ppt and pphm should not be used.

Name	*Symbol*	*Value*	*Examples*
part per hundred	pph	10^{-2}	(Exactly equivalent to percent, %)
part per thousand	ppt	10^{-3}	Atmospheric carbon dioxide is depleted in carbon-13 mass fraction by 7‰ (or 7 ppt) relative to ocean water
permille[1]	‰	10^{-3}	
part per hundred million	pphm	10^{-8}	The mass fraction of impurity in the metal was less than 5 pphm
part per billion	ppb	10^{-9}	The air quality standard for ozone is a volume fraction of $\phi = 120$ ppb
part per trillion	ppt	10^{-12}	The natural background volume fraction of NO in air was found to be $\phi = 140$ ppt
part per quadrillion	ppq	10^{-15}	

(1) The permille is also spelled per mille, per mill, permil or pro mille.

Units of logarithmic quantities: neper, bel and decibel

In some fields, especially in acoustics, special names are given to the number 1 when expressing physical quantities defined in terms of the logarithm of a ratio. For a damped linear oscillation the amplitude of a quantity as a function of time is given by

$$F(t) = A\mathrm{e}^{-\delta t}\cos\omega t = A\,\mathrm{Re}\{\exp[(-\delta + \mathrm{i}\omega)t]\}$$

From this relation it is clear that the coherent SI unit for the damping coefficient δ and the angular frequency ω is the reciprocal second (s^{-1}). However, the special names neper, Np, and radian, rad (see p.11 and p.72), are used for the units of the dimensionless products δt and ωt respectively. Similarly the quantities δ and ω may be expressed in the units Np/s and rad/s respectively. Used in this way the neper, Np, and the radian, rad, may both be thought of as special names for the number 1.

In the field of acoustics and signal transmission, signal power levels and signal amplitude levels (or field levels) are usually expressed as the decadic or the napierian logarithm of the ratio of the power P to a reference power P_0, or of the field F to a reference field F_0. Since power is often proportional to the square of the field or amplitude (when the field acts on equal impedances) it is convenient to define the power level and the field level to be equal in such a case. This is done by defining the power level and the field level according to the relations

$$L_F = \ln(F/F_0), \quad \text{and} \quad L_P = \tfrac{1}{2}\ln(P/P_0),$$

so that if $(P/P_0) = (F/F_0)^2$ then $L_P = L_F$. The above equations may be written in the form

$$L_F = \ln(F/F_0)\ \mathrm{Np}, \text{ and } L_P = \tfrac{1}{2}\ln(P/P_0)\ \mathrm{Np}$$

The bel, B, and its more frequently used submultiple the decibel, dB, are used when the field and power levels are calculated using decadic logarithms according to the relations

$$L_P = \lg(P/P_0)\ \mathrm{B} = 10\lg(P/P_0)\ \mathrm{dB},$$

and

$$L_F = 2\lg(F/F_0)\ \mathrm{B} = 20\lg(F/F_0)\ \mathrm{dB}$$

The relation between the bel and the neper follows from comparing these equations with the preceding equations. We obtain

$$L_F = \ln(F/F_0)\ \mathrm{Np} = 2\lg(F/F_0)\ \mathrm{B} = \ln(10)\lg(F/F_0)\ \mathrm{Np}$$

giving

$$\mathrm{B} = 10\ \mathrm{dB} = \tfrac{1}{2}\ln(10)\ \mathrm{Np} = 1.151\,293\ \mathrm{Np}$$

However the bel and the decibel should only be used when expressing power levels as a decadic logarithm, and the neper when expressing field levels using a natural logarithm. In practice the neper and the bel are hardly ever used. Only the decibel is used, to represent the decadic logarithm of a power ratio, particularly in the context of acoustics, and in labelling the controls of power amplifiers. Thus the statement $L_P = n$ dB implies that $10\lg(P/P_0) = n$.

The quantities power level and field level, and the units bel, decibel and neper, are summarized in the table and notes that follow.

Name	*Expression*	*Numerical value × unit*	*Notes*
power level	$L_P = \frac{1}{2}\ln(P/P_0)$	$= \frac{1}{2}\ln(P/P_0)\ \mathrm{Np} = \lg(P/P_0)\ \mathrm{B} = 10\lg(P/P_0)\ \mathrm{dB}$	1–3
field level	$L_F = \ln(F/F_0)$	$= \ln(F/F_0)\ \mathrm{Np} = 2\lg(F/F_0)\ \mathrm{B} = 20\lg(F/F_0)\ \mathrm{dB}$	4–6

(1) P_0 is a reference power, which should be specified. The factor $\frac{1}{2}$ is included in the definition to make $L_P \approx L_F$.
(2) In the context of acoustics the power level is called the sound power level and given the symbol L_W, and the reference power $P_0 = 1$ pW.
(3) For example, when $L_P = 1\ \mathrm{B} = 10\ \mathrm{dB}$, $P/P_0 = 10$; and when $L_P = 2\ \mathrm{B} = 20\ \mathrm{dB}$, $P/P_0 = 100$; etc.
(4) F_0 is a reference field, which should be specified.
(5) In the context of acoustics the field level is called the sound pressure level and given the symbol L_p, and the reference pressure $p_0 = 20\ \mu\mathrm{Pa}$.
(6) For example, when $L_F = 1$ Np, $F/F_0 = \mathrm{e} = 2.718281\ \ldots$.

4
Recommended mathematical symbols

4.1 PRINTING OF NUMBERS AND MATHEMATICAL SYMBOLS [5.a]

(i) Numbers in general should be printed in roman (upright) type. The decimal sign between digits in a number should be a point (e.g. 2.3) or a comma (e.g. 2,3). ISO [5.a] recommends a comma in preference to a point for the decimal marker. To facilitate the reading of long numbers the digits may be grouped in threes about the decimal sign but no point or comma should be used except for the decimal sign. When the decimal sign is placed before the first significant digit of a number a zero should always precede the decimal sign.

Examples 2 573.421 736 or 2 573,421 736 or 0.2573×10^4 or $0{,}2573 \times 10^4$

(ii) Numerical values of physical quantities which have been experimentally determined are usually subject to some uncertainty. The experimental uncertainty should always be specified. The magnitude of the uncertainty may be represented as follows.

Examples $l = (5.3478 \pm 0.0065)$ cm or $l = 5.3478$ cm $\pm$ 0.0065 cm
$l = 5.3478\ (32)$ cm
$l = 5.34_8$ cm

In the first example the range of uncertainty is indicated directly as $a \pm b$. It is recommended that this notation should be used only with the meaning that the interval $a \pm b$ contains the true value with a high degree of certainty, such that $b \geq 2\sigma$, where σ denotes the standard uncertainty or standard deviation.

In the second example, a (b), the range of uncertainty b indicated in parenthesis is assumed to apply to the least significant digits of a. It is recommended that this notation be reserved for the meaning that b represents 1 σ in the final digits of a. The third example implies a less precise estimate of uncertainty, which would be read as between 1 and 9 in the subscripted digit. In any case the convention used for uncertainties should be clearly stated.

(iii) Letter symbols for mathematical constants (e.g. e, π, $\mathrm{i} = \sqrt{-1}$) should be printed in roman (upright) type, but letter symbols for numbers other than constants (e.g. quantum numbers) should be printed in italic (sloping) type, similar to physical quantities.

(iv) Symbols for special mathematical functions (e.g. log, lg, exp, sin, cos, d, δ, Δ, ∇, . . .) should be printed in roman type, but symbols for a general function (e.g. $f(x)$, $F(x, y)$, . . .) should be printed in italic type.

(v) Symbols for symmetry species in group theory (e.g. S, P, D, . . . , s, p, d, . . . , Σ, Π, Δ, . . . , $\mathrm{A_{1g}}$, $\mathrm{B_2''}$, . . .) should be printed in roman (upright) type when they represent the state symbol for an atom or a molecule, although they are often printed in italic type when they represent the symmetry species of a point group.

(vi) Vectors and matrices should be printed in bold face italic type.

Examples force $\boldsymbol{F}$, electric field $\boldsymbol{E}$, vector coordinate $\boldsymbol{r}$

Ordinary italic type is used to denote the magnitude of the corresponding vector.

Example $r = |\boldsymbol{r}|$

Tensor quantities may be printed in bold face italic sans-serif type.

Examples $\boldsymbol{\mathsf{S}}$, $\boldsymbol{\mathsf{T}}$

4.2 SYMBOLS, OPERATORS AND FUNCTIONS [5.m]

equal to	$=$	less than	$<$
not equal to	$\neq$	greater than	$>$
identically equal to	$\equiv$	less than or equal to	$\leqslant$
equal by definition to	$\stackrel{\text{def}}{=}$	greater than or equal to	$\geqslant$
approximately equal to	$\approx$	much less than	$\ll$
asymptotically equal to	$\simeq$	much greater than	$\gg$
corresponds to	$\triangleq$	plus	$+$
proportional to	$\propto$, $\sim$	minus	$-$
tend to, approaches	$\rightarrow$	plus or minus	$\pm$
infinity	∞	minus or plus	$\mp$

a multiplied by b [1]	$a\,b$, ab, $a \cdot b$, $a \times b$
a divided by b	a/b, ab^{-1}, $\frac{a}{b}$
magnitude of a	$\lvert a \rvert$
a to the power n	a^n
square root of a, and of $a^2 + b^2$	$\sqrt{a}$, $a^{1/2}$, $\sqrt{a^2 + b^2}$, $(a^2 + b^2)^{1/2}$
nth root of a	$a^{1/n}$, $\sqrt[n]{a}$,
mean value of a	$\langle a \rangle$, $\bar{a}$
sign of a (equal to $a/\lvert a \rvert$)	$\operatorname{sgn} a$
n factorial	$n!$
binomial coefficient $= n!/p!(n-p)!$	C_p^n, $\binom{n}{p}$
sum of a_i	$\sum a_i$, $\sum_i a_i$, $\sum_{i=1}^{n} a_i$
product of a_i	$\prod a_i$, $\prod_i a_i$, $\prod_{i=1}^{n} a_i$
sine of x	$\sin x$
cosine of x	$\cos x$
tangent of x	$\tan x$
cotangent of x	$\cot x$
inverse sine of x	$\arcsin x$
inverse cosine of x	$\arccos x$
inverse tangent of x	$\arctan x$
hyperbolic sine of x	$\sinh x$
hyperbolic cosine of x	$\cosh x$
hyperbolic tangent of x	$\tanh x$
hyperbolic cotangent of x	$\coth x$
base of natural logarithms	e
exponential of x	$\exp x$, e^x
natural logarithm of x	$\ln x$, $\log_{\mathrm{e}} x$
logarithm to the base a of x	$\log_a x$
logarithm to the base 10 of x	$\lg x$, $\log_{10} x$
logarithm to the base 2 of x	$\operatorname{lb} x$, $\log_2 x$

(1) When multiplication is indicated by a dot, the dot should be raised: $a \cdot b$.

square root of minus one	i
real part of $z = a + \mathrm{i}b$	$\mathrm{Re}\, z = a$
imaginary part of $z = a + \mathrm{i}b$	$\mathrm{Im}\, z = b$
modulus of $z = a + \mathrm{i}b$, absolute value of $z = a + \mathrm{i}b$	$\|z\| = (a^2 + b^2)^{1/2}$
argument of $z = a + \mathrm{i}b$	$\arg z = \arctan(b/a)$
complex conjugate of $z = a + \mathrm{i}b$	$z^* = a - \mathrm{i}b$
greatest integer $\leqslant x$	$\mathrm{ent}\, x$, $\mathrm{int}\, x$
integer division, $\mathrm{ent}(n/m)$	$n \,\mathrm{div}\, m$
remainder after integer division, $n/m - \mathrm{ent}(n/m)$	$n \bmod m$
change in x	$\Delta x = x(\text{final}) - x(\text{initial})$
infinitesimal change of f	δf
limit of $f(x)$ as x tends to a	$\lim\limits_{x \to a} f(x)$
1st derivative of f	$\mathrm{d}f/\mathrm{d}x$, $\partial_x f$, $\mathrm{D}_x f$, f'
nth derivative of f	$\mathrm{d}^n f/\mathrm{d}x^n$, $f'' \cdots$
partial derivative of f	$\partial f/\partial x$
total differential of f	$\mathrm{d}f$
inexact differential of f (note 2)	$\text{đ}f$
first derivative of x with respect to time	$\dot{x}$, $\partial x/\partial t$
integral of $f(x)$	$\int f(x)\,\mathrm{d}x$, $\int \mathrm{d}x\, f(x)$
Kronecker delta	$\delta_{ij} = 1$ if $i = j$, $= 0$ if $i \neq j$
Levi–Civita symbol	$\varepsilon_{ijk} = 1$ if i, j, k is a cyclic permutation, $= -1$ if i, j, k is anticyclic, $= 0$ otherwise.
Dirac delta function (distribution)	$\delta(x)$, $\int f(x)\delta(x)\,\mathrm{d}x = f(0)$
unit step function, Heaviside function	$\varepsilon(x)$, $\mathrm{H}(x)$ $\varepsilon(x) = 1$ for $x > 0$, $= 0$ for $x < 0$
gamma function	$\Gamma(x) = \int t^{x-1} \mathrm{e}^{-t}\,\mathrm{d}t$ $= (x-1)!$ for integer values of x
convolution of functions f and g	$f * g = \int f(x - x')g(x')\,\mathrm{d}x'$

vectors

vector a	$\boldsymbol{a}$, $(\vec{a})$
cartesian components of a	a_x, a_y, a_z
unit vectors in cartesian axes	$\boldsymbol{i}, \boldsymbol{j}, \boldsymbol{k}$, or $\boldsymbol{e}_x$, $\boldsymbol{e}_y$, $\boldsymbol{e}_z$
scalar product	$\boldsymbol{a} \cdot \boldsymbol{b}$
vector or cross product	$\boldsymbol{a} \times \boldsymbol{b}$, $\boldsymbol{a} \wedge \boldsymbol{b}$
nabla operator, del operator	$\nabla = \boldsymbol{i}\partial/\partial x + \boldsymbol{j}\partial/\partial y + \boldsymbol{k}\partial/\partial z$
Laplacian operator	∇^2, $\Delta = \partial^2/\partial x^2 + \partial^2/\partial y^2 + \partial^2/\partial z^2$
gradient of a scalar field V	$\mathrm{grad}\, V$, ∇V
divergence of a vector field $\boldsymbol{A}$	$\mathrm{div}\, \boldsymbol{A}$, $\nabla \cdot \boldsymbol{A}$
curl of a vector field $\boldsymbol{A}$	$\mathrm{curl}\, \boldsymbol{A}$, $\mathrm{rot}\, \boldsymbol{A}$, $\nabla \times \boldsymbol{A}$

matrices

matrix of elements A_{ij}	$\boldsymbol{A}$
product of matrices $\boldsymbol{A}$ and $\boldsymbol{B}$	$\boldsymbol{AB}$, $(\boldsymbol{AB})_{ik} = \sum\limits_j A_{ij}B_{jk}$

(2) Notation used in thermodynamics, see p.48, note (1).

(double) scalar product of $\boldsymbol{A}$ and $\boldsymbol{B}$	$\boldsymbol{A}:\boldsymbol{B} = \sum_{i,j} A_{ij}B_{ji}$
unit matrix	$\boldsymbol{E}, \boldsymbol{I}$
inverse of a square matrix $\boldsymbol{A}$	$\boldsymbol{A}^{-1}$
transpose of matrix $\boldsymbol{A}$	$\boldsymbol{A}^{\mathrm{T}}, \tilde{\boldsymbol{A}}, \boldsymbol{A}'$
complex conjugate of matrix $\boldsymbol{A}$	$\boldsymbol{A}^*$
conjugate transpose of $\boldsymbol{A}$ (hermitian conjugate of $\boldsymbol{A}$)	$\boldsymbol{A}^\dagger, (\boldsymbol{A}^\dagger)_{ij} = A_{ji}{}^*$
trace of square matrix $\boldsymbol{A}$	$\mathrm{tr}\,\boldsymbol{A}, \mathrm{Tr}(\boldsymbol{A}), \Sigma_i A_{ii}$
determinant of square matrix $\boldsymbol{A}$	$\det \boldsymbol{A}, \lvert\boldsymbol{A}\rvert$

logical operators

A is contained in B	$A \subset B$
union of A and B	$A \cup B$
intersection of A and B	$A \cap B$
p and q (conjunction sign)	$p \wedge q$
p or q or both (disjunction sign)	$p \vee q$
x belongs to A	$x \in A$
x does not belong to A	$x \notin A$
the set A contains x	$A \ni x$
difference of A and B	$A \setminus B$

5
Fundamental physical constants

The following values were recommended by the CODATA Task Group on Fundamental Constants in 1986 [70]. For each constant the standard deviation uncertainty in the least significant digits is given in parentheses.

Quantity	*Symbol*	*Value*
permeability of vacuum[1]	μ_0	$4\pi \times 10^{-7}$ H m^{-1} (defined)
speed of light in vacuum	c_0	299 792 458 m s^{-1} (defined)
permittivity of vacuum[1]	$\varepsilon_0 = 1/\mu_0 c_0^2$	8.854 187 816 ... $\times 10^{-12}$ F m^{-1}
Planck constant	h	6.626 075 5(40) $\times 10^{-34}$ J s
	$\hbar = h/2\pi$	1.054 572 66(63) $\times 10^{-34}$ J s
elementary charge	e	1.602 177 33(49) $\times 10^{-19}$ C
electron rest mass,	m_e	9.109 389 7(54) $\times 10^{-31}$ kg
proton rest mass	m_p	1.672 623 1(10) $\times 10^{-27}$ kg
neutron rest mass	m_n	1.674 928 6(10) $\times 10^{-27}$ kg
atomic mass constant, (unified atomic mass unit)	$m_u = 1$ u	1.660 540 2(10) $\times 10^{-27}$ kg
Avogadro constant	L, N_A	6.022 136 7(36) $\times 10^{23}$ mol^{-1}
Boltzmann constant	k	1.380 658(12) $\times 10^{-23}$ J K^{-1}
Faraday constant	F	9.648 530 9(29) $\times 10^{4}$ C mol^{-1}
gas constant	R	8.314 510(70) J K^{-1} mol^{-1}
zero of the Celsius scale		273.15 K (defined)
molar volume, ideal gas, $p = 1$ bar, $\theta = 0\,°C$		22.711 08(19) L mol^{-1}
standard atmosphere	atm	101 325 Pa (defined)
fine structure constant	$\alpha = \mu_0 e^2 c_0/2h$	7.297 353 08(33) $\times 10^{-3}$
	α^{-1}	137.035 989 5(61)
Bohr radius	$a_0 = 4\pi\varepsilon_0\hbar^2/m_e e^2$	5.291 772 49(24) $\times 10^{-11}$ m
Hartree energy	$E_h = \hbar^2/m_e a_0^2$	4.359 748 2(26) $\times 10^{-18}$ J
Rydberg constant	$R_\infty = E_h/2hc_0$	1.097 373 153 4(13) $\times 10^{7}$ m^{-1}
Bohr magneton	$\mu_B = e\hbar/2m_e$	9.274 015 4(31) $\times 10^{-24}$ J T^{-1}
electron magnetic moment	μ_e	9.284 770 1(31) $\times 10^{-24}$ J T^{-1}
Landé g-factor for free electron	$g_e = 2\mu_e/\mu_B$	2.002 319 304 386(20)
nuclear magneton	$\mu_N = (m_e/m_p)\mu_B$	5.050 786 6(17) $\times 10^{-27}$ J T^{-1}
proton magnetic moment	μ_p	1.410 607 61(47) $\times 10^{-26}$ J T^{-1}
proton magnetogyric ratio	γ_p	2.675 221 28(81) $\times 10^{8}$ s^{-1} T^{-1}
magnetic moment of protons in H_2O, μ'_p	μ'_p/μ_B	1.520 993 129(17) $\times 10^{-3}$
proton resonance frequency per field in H_2O	$\gamma'_p/2\pi$	42.576 375(13) MHz T^{-1}
Stefan–Boltzmann constant	$\sigma = 2\pi^5 k^4/15h^3 c_0^2$	5.670 51(19) $\times 10^{-8}$ W m^{-2} K^{-4}
first radiation constant	$c_1 = 2\pi h c_0^2$	3.741 774 9(22) $\times 10^{-16}$ W m^{2}
second radiation constant	$c_2 = hc_0/k$	1.438 769(12) $\times 10^{-2}$ m K
gravitational constant	G	6.672 59(85) $\times 10^{-11}$ m^3 kg^{-1} s^{-2}
standard acceleration of free fall	g_n	9.806 65 m s^{-2} (defined)

(1) H m^{-1} = N A^{-2} = N s^2 C^{-2}; F m^{-1} = C^2 J^{-1} m^{-1}; ε_0 may be calculated exactly from the defined values of μ_0 and c_0.

Values of common mathematical constants

Mathematical constant	*Symbol*	*Value*
ratio of circumference to diameter of a circle[2]	π	3.141 592 653 59
base of natural logarithms	e	2.718 281 828 46
natural logarithm of 10	ln 10	2.302 585 092 99

(2) A mnemonic for π, based on the number of letters in words of the English language, is:

'How I like a drink, alcoholic of course, after the heavy lectures involving quantum mechanics!'

There are similar mnemonics in poem form in French:

'Que j'aime à faire apprendre ce nombre utile aux sages!
Immortel Archimède, artiste ingénieur,
Qui de ton jugement peut priser la valeur?
Pour moi, ton problème eut de pareils avantages.'

and German:

'Wie? O! Dies π
Macht ernstlich so vielen viele Müh'!
Lernt immerhin, Jünglinge, leichte Verselein,
Wie so zum Beispiel dies dürfte zu merken sein!'.

See the Japanese [2.d] and Russian [2.b] editions for further mnemonics.

6
Properties of particles, elements and nuclides

The symbols for particles, chemical elements and nuclides have been discussed in section 2.10. The recently recommended systematic nomenclature and symbolism for chemical elements of atomic number greater than 103 is briefly described in footnote U to table 6.2.

6.1 PROPERTIES OF SOME PARTICLES

The data given in the table are taken from the compilations by Cohen and Taylor [70], the Particle Data Group [71] and by Wapstra and Audi [72].

Name	Symbol[a]	Spin I	Charge number z	Rest mass m/u	Rest mass mc^2/MeV	Magnetic moment μ/μ_N	Mean life τ/s
photon	γ	1	0	0	0		
neutrino	ν_e	1/2	0	0	0		
electron[b]	e	1/2	−1	$5.485\,799\,03(13) \times 10^{-4}$	0.510 999 06 (15)	1.001 159 652 193 (10)[c]	
muon	$\mu^{\pm}$	1/2	±1	0.113 428 913 (17)	105.658 389 (34)	1.001 165 923 (8)[d]	$2.197\,03(4) \times 10^{-6}$
pion	$\pi^{\pm}$	1	±1	0.149 832 3 (8)	139.5679 (7)		$2.6030(24) \times 10^{-8}$
pion	π^0	1	0	0.144 9008 (9)	134.9743 (8)		$8.4(6) \times 10^{-17}$
proton	p	1/2	1	1.007 276 470 (12)	938.272 31 (28)	2.792 847 386 (63)	
neutron	n	1/2	0	1.008 664 904 (14)	939.565 63 (28)	−1.913 042 75 (45)	889.1 (21)
deuteron	d	1	1	2.013 553 214 (24)	1875.613 39 (53)	0.857 437 6 (1)	
triton	t	1/2	1	3.015 500 71 (4)	2808.921 78 (85)	2.978 960 (1)	
helion	h	1/2	2	3.014 932 23 (4)	2808.392 25 (85)	−2.127 624 (1)	
α-particle	α	0	2	4.001 506 170 (50)	3727.380 3 (11)	0	

(a) The Particle Data Group recommends the use of italic symbols for particles and this has been adopted by many physicists [71].
(b) The electron as β-particle is sometimes denoted by β.
(c) The value is given in Bohr magnetons μ/μ_B, $\mu_B = e\hbar/2m_e$.
(d) The value is given as μ/μ_μ where $\mu_\mu = e\hbar/2m_\mu$.

In nuclear physics and chemistry the masses of particles are often quoted as their energy equivalents (usually in megaelectronvolts). The unified atomic mass unit corresponds to 931.494 32 (28) MeV [70].

Atom-like pairs of a positive particle and an electron are sometimes sufficiently stable to be treated as individual entities with special names.

Examples positronium (e^+e^-) $m(e^+e^-) = 1.097\,152\,503(26) \times 10^{-3}$ u

muonium (μ^+e^-; Mu) $m(\text{Mu}) = 0.113\,977\,478(17)$ u

The positive or negative sign for the magnetic moment of a particle implies that the orientation of the magnetic dipole with respect to the angular momentum corresponds to the rotation of a positive or negative charge respectively.

6.2 STANDARD ATOMIC WEIGHTS OF THE ELEMENTS 1991

As agreed by the IUPAC Commission on Atomic Weights and Isotopic Abundances in 1979 [42] the relative atomic mass (atomic weight) of an element, E, can be defined for any specified sample. It is the average mass of its atoms in the sample divided by the unified atomic mass unit[1] or alternatively the molar mass of its atoms divided by the standard molar mass $M^{\ominus} = Lm_u = 1\,\mathrm{g\,mol^{-1}}$:

$$A_r(\mathrm{E}) = \overline{m}_a(\mathrm{E})/\mathrm{u} = M(\mathrm{E})/M^{\ominus}$$

The variations in isotopic composition of many elements in samples of different origin limit the precision to which a relative atomic mass can be given. The standard atomic weights revised biennially by the IUPAC Commission on Atomic Weights and Isotopic Abundances are meant to be applicable for normal materials. This means that to a high level of confidence the relative atomic mass of an element in any normal sample will be within the uncertainty limits of the tabulated value. By 'normal' it is meant here that the material is a reasonably possible source of the element or its compounds in commerce for industry and science and that it has not been subject to significant modification of isotopic composition within a geologically brief period [43]. This, of course, excludes materials studied themselves for very anomalous isotopic composition.

Table 6.2 lists the relative atomic masses of the elements in the alphabetical order of chemical symbols. The values have been recommended by the IUPAC Commission on Atomic Weights and Isotopic Abundances in 1991 [44] and apply to elements as they exist naturally on earth.

The relative atomic masses of many elements depend on the origin and treatment of the materials [45]. The notes to this table explain the types of variation to be expected for individual elements. When used with due regard to the notes the values are considered reliable to ± the figure given in parentheses being applicable to the last digit. For elements without a characteristic terrestrial isotopic composition no standard atomic weight is recommended. The atomic mass of its most stable isotope can be found in table 6.3.

Symbol	*Atomic number*	*Name*	*Relative atomic mass (atomic weight)*	*Note*
Ac	89	actinium		A
Ag	47	silver	107.868 2 (2)	g
Al	13	aluminium	26.981 539 (5)	
Am	95	americium		A
Ar	18	argon	39.948 (1)	g, r
As	33	arsenic	74.921 59 (2)	
At	85	astatine		A
Au	79	gold	196.966 54 (3)	
B	5	boron	10.811 (5)	g, m, r
Ba	56	barium	137.327 (7)	
Be	4	beryllium	9.012 182 (3)	
Bi	83	bismuth	208.980 37 (3)	
Bk	97	berkelium		A
Br	35	bromine	79.904 (1)	
C	6	carbon	12.011 (1)	r

(1) Note that the atomic mass constant, m_u, is equal to the unified atomic mass unit, u, and is defined in terms of the mass of the carbon-12 atom: $m_u = 1\ \mathrm{u} = m_a(^{12}\mathrm{C})/12$.

Symbol	*Atomic number*	*Name*	*Relative atomic mass (atomic weight)*	*Note*
Ca	20	calcium	40.078 (4)	g
Cd	48	cadmium	112.411 (8)	g
Ce	58	cerium	140.115 (4)	g
Cf	98	californium		A
Cl	17	chlorine	35.452 7 (9)	m
Cm	96	curium		A
Co	27	cobalt	58.933 20 (1)	
Cr	24	chromium	51.996 1 (6)	
Cs	55	caesium	132.905 43 (5)	
Cu	29	copper	63.546 (3)	r
Dy	66	dysprosium	162.50 (3)	g
Er	68	erbium	167.26 (3)	g
Es	99	einsteinium		A
Eu	63	europium	151.965 (9)	g
F	9	fluorine	18.998 403 2 (9)	
Fe	26	iron	55.847 (3)	
Fm	100	fermium		A
Fr	87	francium		A
Ga	31	gallium	69.723 (1)	
Gd	64	gadolinium	157.25 (3)	g
Ge	32	germanium	72.61 (2)	
H	1	hydrogen	1.007 94 (7)	g, m, r
He	2	helium	4.002 602 (2)	g, r
Hf	72	hafnium	178.49 (2)	
Hg	80	mercury	200.59 (2)	
Ho	67	holmium	164.930 32 (3)	
I	53	iodine	126.904 47 (3)	
In	49	indium	114.818 (3)	
Ir	77	iridium	192.22 (3)	
K	19	potassium	39.098 3 (1)	
Kr	36	krypton	83.80 (1)	g, m
La	57	lanthanum	138.905 5 (2)	g
Li	3	lithium	6.941 (2)	g, m, r
Lr	103	lawrencium		A
Lu	71	lutetium	174.967 (1)	g
Md	101	mendelevium		A
Mg	12	magnesium	24.305 0 (6)	
Mn	25	manganese	54.938 05 (1)	
Mo	42	molybdenum	95.94 (1)	g
N	7	nitrogen	14.006 74 (7)	g, r
Na	11	sodium	22.989 768 (6)	
Nb	41	niobium	92.906 38 (2)	
Nd	60	neodymium	144.24 (3)	g
Ne	10	neon	20.179 7 (6)	g, m
Ni	28	nickel	58.34 (2)	

Symbol	Atomic number	Name	Relative atomic mass (atomic weight)	Note
No	102	nobelium		A
Np	93	neptunium		A
O	8	oxygen	15.999 4 (3)	g, r
Os	76	osmium	190.23 (3)	g
P	15	phosphorus	30.973 762 (4)	
Pa	91	protactinium	231.035 88 (2)	Z
Pb	82	lead	207.2 (1)	g, r
Pd	46	palladium	106.42 (1)	g
Pm	61	promethium		A
Po	84	polonium		A
Pr	59	praseodymium	140.907 65 (3)	
Pt	78	platinum	195.08 (3)	
Pu	94	plutonium		A
Ra	88	radium		A
Rb	37	rubidium	85.467 8 (3)	g
Re	75	rhenium	186.207 (1)	
Rh	45	rhodium	102.905 50 (3)	
Rn	86	radon		A
Ru	44	ruthenium	101.07 (2)	g
S	16	sulfur	32.066 (6)	g, r
Sb	51	antimony	121.757 (3)	g
Sc	21	scandium	44.955 910 (9)	
Se	34	selenium	78.96 (3)	
Si	14	silicon	28.085 5 (3)	r
Sm	62	samarium	150.36 (3)	g
Sn	50	tin	118.710 (7)	g
Sr	38	strontium	87.62 (1)	g, r
Ta	73	tantalum	180.947 9 (1)	
Tb	65	terbium	158.925 34 (3)	
Tc	43	technetium		A
Te	52	tellurium	127.60 (3)	g
Th	90	thorium	232.038 1 (1)	g, Z
Ti	22	titanium	47.88 (3)	
Tl	81	thallium	204.383 3 (2)	
Tm	69	thulium	168.934 21 (3)	
U	92	uranium	238.028 9 (1)	g, m, Z
Une	109	unnilennium		A, U
Unh	106	unnilhexium		A, U
Uno	108	unniloctium		A, U
Unp	105	unnilpentium		A, U
Unq	104	unnilquadium		A, U
Uns	107	unnilseptium		A, U
V	23	vanadium	50.941 5 (1)	
W	74	tungsten	183.84 (1)	
Xe	54	xenon	131.29 (2)	g, m
Y	39	yttrium	88.905 85 (2)	

Symbol	*Atomic number*	*Name*	*Relative atomic mass (atomic weight)*	*Note*
Yb	70	ytterbium	173.04(3)	g
Zn	30	zinc	65.39(2)	
Zr	40	zirconium	91.224(2)	g

(g) **g**eologically exceptional specimens are known in which the element has an isotopic composition outside the limits for normal material. The difference between the average relative atomic mass of the element in such specimens and that given in the table may exceed considerably the implied uncertainty.
(m) **m**odified isotopic compositions may be found in commercially available material because it has been subjected to an undisclosed or inadvertent isotopic separation. Substantial deviations in relative atomic mass of the element from that given in the table can occur.
(r) **r**ange in isotopic composition of normal terrestrial material prevents a more precise relative atomic mass being given; the tabulated $A_r(\mathrm{E})$ value should be applicable to any normal material.
(A) Radioactive element that lacks a characteristic terrestrial isotopic composition.
(Z) An element without stable nuclide(s), exhibiting a range of characteristic terrestrial compositions of long-lived radionuclide(s) such that a meaningful relative atomic mass can be given.
(U) The names and symbols given here are systematic and based on the atomic numbers of the elements as recommended by the IUPAC Commission on the Nomenclature of Inorganic Chemistry [22]. The names are composed of the following roots representing digits of the atomic number:

1 un,	2 bi,	3 tri,	4 quad,	5 pent,
6 hex,	7 sept,	8 oct,	9 enn,	0 nil

The ending -ium is then added to the three roots. The three-letter symbols are derived from the first letters of the corresponding roots.

6.3 PROPERTIES OF NUCLIDES

The table contains the following properties of naturally occurring and some unstable nuclides:

Column
1. Z is the atomic number (number of protons) of the nuclide.
2. Symbol of the element.
3. A is the mass number of the nuclide. The * sign denotes an unstable nuclide (for elements without naturally occurring isotopes it is the most stable nuclide) and the # sign a nuclide of sufficiently long lifetime to enable the determination of its isotopic abundance.
4. The atomic mass is given in unified atomic mass units, $u = m_a(^{12}C)/12$, together with the standard errors in parentheses and applicable to the last digits quoted. The data were extracted from a more extensive list of *The 1983 Atomic Mass Evaluation* by Wapstra and Audi [72].
5. Isotopic abundances are given as mole fractions, x, of the corresponding atoms in percents. They were recommended in 1989 by the IUPAC Commission on Atomic Weights and Isotopic Abundances [45] and are consistent with the standard atomic weights given in table 6.2. The uncertainties given in parentheses are applicable to the last digits quoted and cover the range of probable variations in the materials as well as experimental errors.
6. I is the nuclear spin quantum number.
7. Under magnetic moment the maximum z-component expectation value of the magnetic dipole moment, m, in nuclear magnetons is given. The positive or negative sign implies that the orientation of the magnetic dipole with respect to the angular momentum corresponds to the rotation of a positive or negative charge, respectively. The data were extracted from the compilation by P. Raghavan [73]. An asterisk * indicates that more than one value is given in the original compilation. The value of highest precision or most recent date is given here.
8. Under quadrupole moment, the electric quadrupole moment area (see note 12 on p. 21) is given in units of square femtometres, $fm^2 = 10^{-30}\ m^2$, although most of the tables quote them in barns ($1\ barn = 10^{-28}\ m^2 = 100\ fm^2$). The positive sign implies a prolate nucleus, the negative sign an oblate nucleus. The data for $Z \leq 20$ were taken from the compilation by P. Pyykkö [74] with values for Cl and Ca corrected by D. Sundholm (private communication), and the others from P. Raghavan [73]. An asterisk * indicates that more than one value is given in the original compilation.

Z	Symbol	A	Atomic mass, m_a/u	Isotopic abundance, $100\ x$	Nuclear spin, I	Magnetic moment, m/μ_N	Quadrupole moment, Q/fm^2
1	H	1	1.007 825 035 (12)	99.985 (1)	1/2	+2.792 847 386 (63)	
	(D)	2	2.014 101 779 (24)	0.015 (1)	1	+0.857 438 230 (24)	+0.2860 (15)
	(T)	3*	3.016 049 27 (4)		1/2	+2.978 962 479 (68)	
2	He	3	3.016 029 31 (4)	0.000 137 (3)	1/2	−2.127 624 848 (66)	
		4	4.002 603 24 (5)	99.999 863 (3)	0	0	
3	Li	6	6.015 1214 (7)	7.5 (2)	1	+0.822 056 67 (26)*	−0.082 (4)
		7	7.016 0030 (9)	92.5 (2)	3/2	+3.256 462 53 (40)*	−4.01
4	Be	9	9.012 1822 (4)	100	3/2	−1.177 492 (17)*	+5.288 (38)
5	B	10	10.012 9369 (3)	19.9 (2)	3	+1.800 644 75 (57)	+8.459 (24)
		11	11.009 3054 (4)	80.1 (2)	3/2	+2.688 6489 (10)	+4.059 (10)
6	C	12	12 (by definition)	98.90 (3)	0	0	
		13	13.003 354 826 (17)	1.10 (3)	1/2	+0.702 4118 (14)	
		14*	14.003 241 982 (27)		0	0	

Z	*Symbol*	A	*Atomic mass,* m_a/u	*Isotopic abundance,* $100x$	*Nuclear spin,* I	*Magnetic moment,* m/μ_N	*Quadrupole moment,* Q/fm^2
7	N	14	14.003 074 002 (26)	99.634 (9)	1	+0.403 761 00 (6)	+2.01 (2)
		15	15.000 108 97 (4)	0.366 (9)	1/2	−0.283 188 842 (45)	
8	O	16	15.994 914 63 (5)	99.762 (15)	0	0	
		17	16.999 1312 (4)	0.038 (3)	5/2	−1.893 80	−2.558 (22)
		18	17.999 1603 (9)	0.200 (12)	0	0	
9	F	19	18.998 403 22 (15)	100	1/2	+2.628 868 (8)	
10	Ne	20	19.992 4356 (22)	90.48 (3)	0	0	
		21	20.993 8428 (21)	0.27 (1)	3/2	−0.661 797 (5)	+10.155 (75)
		22	21.991 3831 (18)	9.25 (3)	0	0	
11	Na	23	22.989 7677 (10)	100	3/2	+2.217 6556 (6)*	+10.06 (20)
12	Mg	24	23.985 0423 (8)	78.99 (3)	0	0	
		25	24.985 8374 (8)	10.00 (1)	5/2	−0.855 465 (8)	+19.94 (20)
		26	25.982 5937 (8)	11.01 (2)	0	0	
13	Al	27	26.981 5386 (8)	100	5/2	+3.641 504 687 (65)	+14.03 (10)
14	Si	28	27.976 9271 (7)	92.23 (1)	0	0	
		29	28.976 4949 (7)	4.67 (1)	1/2	−0.555 29 (3)	
		30	29.973 7707 (7)	3.10 (1)	0	0	
15	P	31	30.973 7620 (6)	100	1/2	+1.131 60 (3)	
16	S	32	31.972 070 70 (25)	95.02 (9)	0	0	
		33	32.971 458 43 (23)	0.75 (1)	3/2	+0.643 8212 (14)	−6.78 (13)
		34	33.967 866 65 (22)	4.21 (8)	0	0	
		36	35.967 080 62 (27)	0.02 (1)	0	0	
17	Cl	35	34.968 852 721 (69)	75.77 (5)	3/2	+0.821 8743 (4)	−8.11 (8)
		37	36.965 902 62 (11)	24.23 (5)	3/2	+0.684 1236 (4)	−6.39 (6)
18	Ar	36	35.967 545 52 (29)	0.337 (3)	0	0	
		38	37.962 7325 (9)	0.063 (1)	0	0	
		40	39.962 3837 (14)	99.600 (3)	0	0	
19	K	39	38.963 7074 (12)	93.2581 (44)	3/2	+0.391 507 31(12)*	+5.9 (6)
		40	39.963 9992 (12)	0.0117 (1)	4	−1.298 1003 (34)	−7.3 (7)
		41	40.961 8254 (12)	6.7302 (44)	3/2	+0.214 870 09 (22)	+7.2 (7)
20	Ca	40	39.962 5906 (13)	96.941 (18)	0	0	
		42	41.958 6176 (13)	0.647 (9)	0	0	
		43	42.958 7662 (13)	0.135 (6)	7/2	−1.317 643 (7)	−4.09 (8)
		44	43.955 4806 (14)	2.086 (12)	0	0	
		46	45.953 689 (4)	0.004 (4)	0	0	
		48	47.952 533 (4)	0.187 (4)	0	0	
21	Sc	45	44.955 9100 (14)	100	7/2	+4.756 4866 (18)	−22 (1)*
22	Ti	46	45.952 6294 (14)	8.0 (1)	0	0	
		47	46.951 7640 (11)	7.3 (1)	5/2	−0.788 48 (1)	+29 (1)
		48	47.947 9473 (11)	73.8 (1)	0	0	
		49	48.947 8711 (11)	5.5 (1)	7/2	−1.104 17 (1)	+24 (1)
		50	49.944 7921 (12)	5.4 (1)	0	0	
23	V	50#	49.947 1609 (17)	0.250 (2)	6	+3.345 6889 (14)	20.9 (40)*
		51	50.943 9617 (17)	99.750 (2)	7/2	+5.148 705 73 (18)	−5.2 (10)*
24	Cr	50	49.946 0464 (17)	4.345 (13)	0	0	
		52	51.940 5098 (17)	83.789 (18)	0	0	
		53	52.940 6513 (17)	9.501 (17)	3/2	−0.474 54 (3)	−15 (5)*
		54	53.938 8825 (17)	2.365 (7)	0	0	
25	Mn	55	54.938 047 1 (16)	100	5/2	+3.468 7190 (9)	+33 (1)*

Z	Symbol	A	Atomic mass, m_a/u	Isotopic abundance, 100x	Nuclear spin, I	Magnetic moment, m/μ_N	Quadrupole moment, Q/fm^2
26	Fe	54	53.939 6127 (15)	5.8 (1)	0	0	
		56	55.934 9393 (16)	91.72 (30)	0	0	
		57	56.935 3958 (16)	2.2 (1)	1/2	+0.090 623 00 (9)*	
		58	57.933 2773 (16)	0.28 (1)	0	0	
27	Co	59	58.933 1976 (16)	100	7/2	+4.627 (9)	+40.4 (40)*
28	Ni	58	57.935 3462 (16)	68.077 (9)	0	0	
		60	59.930 7884 (16)	26.223 (8)	0	0	
		61	60.931 0579 (16)	1.140 (1)	3/2	−0.750 02 (4)	+16.2 (15)
		62	61.928 3461 (16)	3.634 (2)	0	0	
		64	63.927 9679 (17)	0.926 (1)	0	0	
29	Cu	63	62.929 5989 (17)	69.17 (3)	3/2	+2.227 3456 (14)*	−21.1 (4)*
		65	64.927 7929 (20)	30.83 (3)	3/2	+2.381 61 (19)*	−19.5 (4)
30	Zn	64	63.929 1448 (19)	48.6 (3)	0	0	
		66	65.926 0347 (17)	27.9 (2)	0	0	
		67	66.927 1291 (17)	4.1 (1)	5/2	+0.875 2049 (11)*	+15.0 (15)
		68	67.924 8459 (18)	18.8 (4)	0	0	
		70	69.925 325 (4)	0.6 (1)	0	0	
31	Ga	69	68.925 580 (3)	60.108 (9)	3/2	+2.016 589 (44)	+16.8*
		71	70.924 7005 (25)	39.892 (9)	3/2	+2.562 266 (18)	+10.6*
32	Ge	70	69.924 2497 (16)	21.23 (4)	0	0	
		72	71.992 0789 (16)	27.66 (3)	0	0	
		73	72.923 4626 (16)	7.73 (1)	9/2	−0.879 4677 (2)	−17.3 (26)
		74	73.921 1774 (15)	35.94 (2)	0	0	
		76	75.921 4016 (17)	7.44 (2)	0	0	
33	As	75	74.921 5942 (17)	100	3/2	+1.439 475 (65)	+31.4 (6)*
34	Se	74	73.922 4746 (16)	0.89 (2)	0	0	
		76	75.919 2120 (16)	9.36 (1)	0	0	
		77	76.919 9125 (16)	7.63 (6)	1/2	+0.535 074 24 (28)*	
		78	77.917 3076 (16)	23.78 (9)	0	0	
		80	79.916 5196 (19)	49.61 (10)	0	0	
		82	81.916 6978 (23)	8.73 (6)	0	0	
35	Br	79	78.918 3361 (26)	50.69 (7)	3/2	+2.106 400 (4)	+33.1 (4)
		81	80.916 289 (6)	49.31 (7)	3/2	+2.270 562 (4)	+27.6 (4)
36	Kr	78	77.920 396 (9)	0.35 (2)	0	0	
		80	79.916 380 (9)	2.25 (2)	0	0	
		82	81.913 482 (6)	11.6 (1)	0	0	
		83	82.914 135 (4)	11.5 (1)	9/2	−0.970 669 (3)	+25.3 (5)
		84	83.911 507 (4)	57.0 (3)	0	0	
		86	85.910 616 (5)	17.3 (2)	0	0	
37	Rb	85	84.911 794 (3)	72.165 (20)	5/2	+1.353 3515 (8)*	+22.8 (43)*
		87#	86.909 187 (3)	27.835 (20)	3/2	+2.751 818 (2)	+13.2 (1)
38	Sr	84	83.913 430 (4)	0.56 (1)	0	0	
		86	85.909 2672 (28)	9.86 (1)	0	0	
		87	86.908 8841 (28)	7.00 (1)	9/2	−1.093 6030 (13)*	+33.5 (20)
		88	87.905 6188 (28)	82.58 (1)	0	0	
39	Y	89	88.905 849 (3)	100	1/2	−0.137 415 42 (34)*	
40	Zr	90	89.904 7026 (26)	51.45 (3)	0	0	
		91	90.905 6439 (26)	11.22 (4)	5/2	−1.303 62 (2)	−20.6 (10)
		92	91.905 0386 (26)	17.15 (2)	0	0	
		94	93.906 3148 (28)	17.38 (4)	0	0	
		96	95.908 275 (4)	2.80 (2)	0	0	
41	Nb	93	92.906 3772 (27)	100	9/2	+6.1705 (3)	−32 (2)*

Z	Symbol	A	Atomic mass, m_a/u	Isotopic abundance, 100 x	Nuclear spin, I	Magnetic moment, m/μ_N	Quadrupole moment, Q/fm^2
42	Mo	92	91.906 809 (4)	14.84 (4)	0	0	
		94	93.905 0853 (26)	9.25 (3)	0	0	
		95	94.905 8411 (22)	15.92 (5)	5/2	−0.9142 (1)	−2.2 (1)*
		96	95.904 6785 (22)	16.68 (5)	0	0	
		97	96.906 0205 (22)	9.55 (3)	5/2	−0.9335 (1)	+25.5 (13)*
		98	97.905 4073 (22)	24.13 (7)	0	0	
		100	99.907 477 (6)	9.63 (3)	0	0	
43	Tc	98*	97.907 215 (4)		6		
44	Ru	96	95.907 599 (8)	5.52 (6)	0	0	
		98	97.905 287 (7)	1.88 (6)	0	0	
		99	98.905 9389 (23)	12.7 (1)	5/2	−0.6413 (51)*	+7.9 (4)
		100	99.904 2192 (24)	12.6 (1)	0	0	
		101	100.905 5819 (24)	17.0 (1)	5/2	−0.7188 (60)*	+45.7 (23)
		102	101.904 3485 (25)	31.6 (2)	0	0	
		104	103.905 424 (6)	18.7 (2)	0	0	
45	Rh	103	102.905 500 (4)	100	1/2	−0.088 40 (2)	
46	Pd	102	101.905 634 (5)	1.02 (1)	0	0	
		104	103.904 029 (6)	11.14 (8)	0	0	
		105	104.905 079 (6)	22.33 (8)	5/2	−0.642 (3)	+66.0 (11)*
		106	105.903 478 (6)	27.33 (3)	0	0	
		108	107.903 895 (4)	26.46 (9)	0	0	
		110	109.905 167 (20)	11.72 (9)	0	0	
47	Ag	107	106.905 092 (6)	51.839 (7)	1/2	−0.113 679 65 (15)*	
		109	108.904 756 (4)	48.161 (7)	1/2	−0.130 690 62 (22)*	
48	Cd	106	105.906 461 (7)	1.25 (4)	0	0	
		108	107.904 176 (6)	0.89 (2)	0	0	
		110	109.903 005 (4)	12.49 (12)	0	0	
		111	110.904 182 (3)	12.80 (8)	1/2	−0.594 886 07 (84)*	
		112	111.902 757 (3)	24.13 (28)	0	0	
		113 #	112.904 400 (3)	12.22 (8)	1/2	−0.622 300 92 (87)	
		114	113.903 357 (3)	28.73 (28)	0	0	
		116	115.904 755 (4)	7.49 (12)	0	0	
49	In	113	112.904 061 (4)	4.3 (2)	9/2	+5.5289 (2)	+79.9
		115 #	114.903 882 (4)	95.7 (2)	9/2	+5.5408 (2)	+81.0*
50	Sn	112	111.904 826 (5)	0.97 (1)	0	0	
		114	113.902 784 (4)	0.65 (1)	0	0	
		115	114.903 348 (3)	0.34 (1)	1/2	−0.918 83 (7)	
		116	115.901 747 (3)	14.53 (11)	0	0	
		117	116.902 956 (3)	7.68 (7)	1/2	−1.001 04 (7)	
		118	117.901 609 (3)	24.23 (11)	0	0	
		119	118.903 311 (3)	8.59 (4)	1/2	−1.047 28 (7)	
		120	119.902 1991 (29)	32.59 (10)	0	0	
		122	121.903 4404 (30)	4.63 (3)	0	0	
		124	123.905 2743 (17)	5.79 (5)	0	0	
51	Sb	121	120.903 8212 (29)	57.36 (8)	5/2	+3.3634 (3)	−36 (4)*
		123	122.904 2160 (24)	42.64 (8)	7/2	+2.5498 (2)	−49 (5)
52	Te	120	119.904 048 (21)	0.096 (2)	0	0	
		122	121.903 050 (3)	2.603 (4)	0	0	
		123	122.904 2710 (22)	0.908 (2)	1/2	−0.736 9478 (8)	
		124	123.902 8180 (18)	4.816 (6)	0	0	
		125	124.904 4285 (25)	7.139 (6)	1/2	−0.888 505 13 (43)*	
		126	125.903 3095 (25)	18.95 (1)	0	0	
		128	127.904 463 (4)	31.69 (1)	0	0	
		130	129.906 229 (5)	33.80 (1)	0	0	
53	I	127	126.904 473 (5)	100	5/2	+2.813 273 (84)	−78.9

Z	Symbol	A	Atomic mass, m_a/u	Isotopic abundance, 100 x	Nuclear spin, I	Magnetic moment, m/μ_N	Quadrupole moment, Q/fm^2
54	Xe	124	123.905 8942 (22)	0.10 (1)	0	0	
		126	125.904 281 (8)	0.09 (1)	0	0	
		128	127.903 5312 (17)	1.91 (3)	0	0	
		129	128.904 7801 (21)	26.4 (6)	1/2	−0.777 9763 (84)	
		130	129.903 5094 (17)	4.1 (1)	0	0	
		131	130.905 072 (5)	21.2 (4)	3/2	+0.691 8619 (39)	−12.0 (12)
		132	131.904 144 (5)	26.9 (5)	0	0	
		134	133.905 395 (8)	10.4 (2)	0	0	
		136	135.907 214 (8)	8.9 (1)	0	0	
55	Cs	133	132.905 429 (7)	100	7/2	+2.582 0246 (34)*	−0.371 (14)*
56	Ba	130	129.906 282 (8)	0.106 (2)	0	0	
		132	131.905 042 (9)	0.101 (2)	0	0	
		134	133.904 486 (7)	2.417 (27)	0	0	
		135	134.905 665 (7)	6.592 (18)	3/2	+0.837 943 (17)*	+16.0 (3)*
		136	135.904 553 (7)	7.854 (36)	0	0	
		137	136.905 812 (6)	11.23 (4)	3/2	+0.937 365 (20)*	+24.5 (4)*
		138	137.905 232 (6)	71.70 (7)	0	0	
57	La	138 #	137.907 105 (6)	0.0902 (2)	5	+3.713 646 (7)	+45 (2)*
		139	138.906 347 (5)	99.9098 (2)	7/2	+2.783 0455 (9)	+20 (1)
58	Ce	136	135.907 140 (50)	0.19 (1)	0	0	
		138	137.905 985 (12)	0.25 (1)	0	0	
		140	139.905 433 (4)	88.48 (10)	0	0	
		142	141.909 241 (4)	11.08 (10)	0	0	
59	Pr	141	140.907 647 (4)	100	5/2	+4.2754 (5)	−5.89 (42)
60	Nd	142	141.907 719 (4)	27.13 (12)	0	0	
		143	142.909 810 (4)	12.18 (6)	7/2	−1.065 (5)	−63 (6)
		144	143.910 083 (4)	23.80 (12)	0	0	
		145	144.912 570 (4)	8.30 (6)	7/2	−0.656 (4)	−33 (3)
		146	145.913 113 (4)	17.19 (9)	0	0	
		148	147.916 889 (4)	5.76 (3)	0	0	
		150	149.920 887 (4)	5.64 (3)	0	0	
61	Pm	145*	144.912 743 (4)		5/2		
62	Sm	144	143.911 998 (4)	3.1 (1)	0	0	
		147 #	146.914 894 (4)	15.0 (2)	7/2	−0.8148 (7)	−25.9 (26)
		148	147.914 819 (4)	11.3 (1)	0	0	
		149	148.917 180 (4)	13.8 (1)	7/2	−0.6717 (7)*	+7.5 (8)*
		150	149.917 273 (4)	7.4 (1)	0	0	
		152	151.919 728 (4)	26.7 (2)	0	0	
		154	153.922 205 (4)	22.7 (2)	0	0	
63	Eu	151	150.919 702 (8)	47.8 (15)	5/2	+3.4717 (6)	+90.3 (10)*
		153	152.921 225 (4)	52.2 (15)	5/2	+1.5330 (8)*	+241.2 (21)*
64	Gd	152	151.919 786 (4)	0.20 (1)	0	0	
		154	153.920 861 (4)	2.18 (3)	0	0	
		155	154.922 618 (4)	14.80 (5)	3/2	−0.257 23 (35)*	+130 (2)*
		156	155.922 118 (4)	20.47 (4)	0	0	
		157	156.923 956 (4)	15.65 (3)	3/2	−0.337 26 (55)*	+136 (2)*
		158	157.924 019 (4)	24.84 (12)	0	0	
		160	159.927 049 (4)	21.86 (4)	0	0	
65	Tb	159	158.925 342 (4)	100	3/2	+2.014 (4)	+143.2 (8)
66	Dy	156	155.924 277 (8)	0.06 (1)	0	0	
		158	157.924 403 (5)	0.10 (1)	0	0	
		160	159.925 193 (4)	2.34 (6)	0	0	
		161	160.926 930 (4)	18.9 (2)	5/2	−0.4803 (25)*	+250.7 (20)*
		162	161.926 795 (4)	25.5 (2)	0	0	

Z	Symbol	A	Atomic mass, m_a/u	Isotopic abundance, 100 x	Nuclear spin, I	Magnetic moment, m/μ_N	Quadrupole moment, Q/fm^2
66	Dy	163	162.928 728 (4)	24.9 (2)	5/2	+0.6726 (35)	+264.8 (21)
		164	163.929 171 (4)	28.2 (2)	0	0	
67	Ho	165	164.930 319 (4)	100	7/2	+4.173 (27)	+349 (3)*
68	Er	162	161.928 775 (4)	0.14 (1)	0	0	
		164	163.929 198 (4)	1.61 (1)	0	0	
		166	165.930 290 (4)	33.6 (2)	0	0	
		167	166.932 046 (4)	22.95 (15)	7/2	−0.563 85 (12)	+356.5 (29)
		168	167.932 368 (4)	26.8 (2)	0	0	
		170	169.935 461 (4)	14.9 (2)	0	0	
69	Tm	169	168.934 212 (4)	100	1/2	−0.2316 (15)	
70	Yb	168	167.933 894 (5)	0.13 (1)	0	0	
		170	169.934 759 (4)	3.05 (6)	0	0	
		171	170.936 323 (3)	14.3 (2)	1/2	+0.493 67 (1)*	
		172	171.936 378 (3)	21.9 (3)	0	0	
		173	172.938 208 (3)	16.12 (21)	5/2	−0.679 89 (3)*	+280 (4)
		174	173.938 859 (3)	31.8 (4)	0	0	
		176	175.942 564 (4)	12.7 (2)	0	0	
71	Lu	175	174.940 770 (3)	97.41 (2)	7/2	+2.2327 (11)*	+349 (2)*
		176#	175.942 679 (3)	2.59 (2)	7	+3.1692 (45)*	+492 (3)*
72	Hf	174	173.940 044 (4)	0.162 (3)	0	0	
		176	175.941 406 (4)	5.206 (5)	0	0	
		177	176.943 217 (3)	18.606 (4)	7/2	+0.7935 (6)	+336.5 (29)*
		178	177.943 696 (3)	27.297 (4)	0	0	
		179	178.945 8122 (29)	13.629 (6)	9/2	−0.6409 (13)	+379.3 (33)*
		180	179.946 5457 (30)	35.100 (7)	0	0	
73	Ta	180	179.947 462 (4)	0.012 (2)	8		
		181	180.947 992 (3)	99.988 (2)	7/2	+2.3705 (7)	+328 (6)*
74	W	180	179.946 701 (5)	0.13 (4)	0	0	
		182	181.948 202 (3)	26.3 (2)	0	0	
		183	182.950 220 (3)	14.3 (1)	1/2	+0.117 784 76 (9)	
		184	183.950 928 (3)	30.67 (15)	0	0	
		186	185.954 357 (4)	28.6 (2)	0	0	
75	Re	185	184.952 951 (3)	37.40 (2)	5/2	+3.1871 (3)	+218 (2)*
		187#	186.955 744 (3)	62.60 (2)	5/2	+3.2197 (3)	+207 (2)*
76	Os	184	183.952 488 (4)	0.02 (1)	0	0	
		186	185.953 830 (4)	1.58 (30)	0	0	
		187	186.955 741 (3)	1.6 (3)	1/2	+0.064 651 89 (6)	
		188	187.955 830 (3)	13.3 (7)	0	0	
		189	188.958 137 (4)	16.1 (8)	3/2	+0.659 933 (4)	+85.6 (28)
		190	189.958 436 (4)	26.4 (12)	0	0	
		192	191.961 467 (4)	41.0 (8)	0	0	
77	Ir	191	190.960 584 (4)	37.3 (5)	3/2	+0.1507 (6)*	+81.6 (9)*
		193	192.962 917 (4)	62.7 (5)	3/2	+0.1637 (6)*	+75.1 (9)*
78	Pt	190	189.959 917 (7)	0.01 (1)	0	0	
		192	191.961 019 (5)	0.79 (6)	0	0	
		194	193.962 655 (4)	32.9 (6)	0	0	
		195	194.964 766 (4)	33.8 (6)	1/2	+0.609 52 (6)	
		196	195.964 926 (4)	25.3 (6)	0	0	
		198	197.967 869 (6)	7.2 (2)	0	0	
79	Au	197	196.966 543 (4)	100	3/2	+0.148 158 (8)*	+54.7 (16)*
80	Hg	196	195.965 807 (5)	0.15 (1)	0	0	
		198	197.966 743 (4)	9.97 (8)	0	0	

Z	Symbol	A	Atomic mass, m_a/u	Isotopic abundance, 100 x	Nuclear spin, I	Magnetic moment, m/μ_N	Quadrupole moment, Q/fm^2
80	Hg	199	198.968 254 (4)	16.87 (10)	1/2	+0.505 885 49 (85)	
		200	199.968 300 (4)	23.10 (16)	0	0	
		201	200.970 277 (4)	13.18 (8)	3/2	−0.560 2257 (14)*	+38.5 (40)*
		202	201.970 617 (4)	29.86 (20)	0	0	
		204	203.973 467 (5)	6.87 (4)	0	0	
81	Tl	203	202.972 320 (5)	29.524 (14)	1/2	+1.622 257 87 (12)	
		205	204.974 401 (5)	70.476 (14)	1/2	+1.638 214 61 (12)	
82	Pb	204	203.973 020 (5)	1.4 (1)	0	0	
		206	205.974 440 (4)	24.1 (1)	0	0	
		207	206.975 872 (4)	22.1 (1)	1/2	+0.582 583 (9)*	
		208	207.976 627 (4)	52.4 (1)	0	0	
83	Bi	209	208.980 374 (5)	100	9/2	+4.1106 (2)	−37.0 (26)*
84	Po	209*	208.982 404 (5)		1/2		
85	At	210*	209.987 126 (12)				
86	Rn	222*	222.017 571 (3)		0	0	
87	Fr	223*	223.019 733 (4)		3/2	+1.17 (2)	+117 (1)
88	Ra	226*	226.025 403 (3)		0	0	
89	Ac	227*	227.027 750 (3)		3/2	+1.1 (1)	+170 (20)
90	Th	232#	232.038 0508 (23)	100	0	0	
91	Pa	231*	231.035 880 (3)		3/2	2.01 (2)	−172 (5)
92	U	233*	233.039 628 (3)		5/2	0.59 (5)	+366.3 (8)
		234#	234.040 9468 (24)	0.0055 (5)	0	0	
		235#	235.043 9242 (24)	0.7200 (12)	7/2	−0.38 (3)*	+455 (9)*
		238#	238.050 7847 (23)	99.2745 (60)	0	0	
93	Np	237*	237.048 1678 (23)		5/2	+3.14 (4)	+388.6 (6)
94	Pu	244*	244.064 199 (5)		0		
95	Am	243*	243.061 375 (3)		5/2	+1.61 (4)	+420 (130)
96	Cm	247*	247.070 347 (5)				
97	Bk	247*	247.070 300 (6)				
98	Cf	251*	251.079 580 (5)				
99	Es	252*	252.082 944 (23)				
100	Fm	257*	257.095 099 (8)				
101	Md	258*	258.098 57 (22)				
102	No	259*	259.100 931 (12)				
103	Lr	260*	260.105 320 (60)				
104	Unq	261*	261.108 69 (22)				
105	Unp	262*	262.113 76 (16)				
106	Unh	263*	263.118 22 (13)				
107	Uns	262*	262.122 93 (45)				
108	Uno	265*	265.130 16 (99)				
109	Une	266*	266.137 64 (45)				

7
Conversion of units

SI units are recommended for use throughout science and technology. However, some non-SI units are in use, and in a few cases they are likely to remain so for many years. Moreover, the published literature of science makes widespread use of non-SI units. It is thus often necessary to convert the values of physical quantities between SI and other units. This chapter is concerned with facilitating this process.

Section 7.1 gives examples illustrating the use of quantity calculus for converting the values of physical quantities between different units. The table in section 7.2 lists a variety of non-SI units used in chemistry, with the conversion factors to the corresponding SI units. Conversion factors for energy and energy-related units (wavenumber, frequency, temperature and molar energy), and for pressure units, are also presented in tables inside the back cover.

Many of the difficulties in converting units between different systems are associated either with the electromagnetic units, or with atomic units and their relationship to the electromagnetic units. In sections 7.3 and 7.4 the relations involving electromagnetic and atomic units are developed in greater detail to provide a background for the conversion factors presented in the table in section 7.2.

7.1 THE USE OF QUANTITY CALCULUS

Quantity calculus is a system of algebra in which symbols are consistently used to represent physical quantities rather, than their measures, i.e. numerical values in certain units. Thus we always take the values of physical quantities to be the product of a numerical value and a unit (see section 1.1), and we manipulate the symbols for physical quantities, numerical values, and units by the ordinary rules of algebra.[1] This system is recommended for general use in science. Quantity calculus has particular advantages in facilitating the problems of converting between different units and different systems of units, as illustrated by the examples below. In all of these examples the numerical values are approximate.

Example 1. The wavelength λ of one of the yellow lines of sodium is given by

$$\lambda = 5.896 \times 10^{-7}\ \mathrm{m}, \quad \text{or} \quad \lambda/\mathrm{m} = 5.896 \times 10^{-7}$$

The ångström is defined by the equation (see table 7.2, under length)

$$1\ \text{Å} = \text{Å} = 10^{-10}\ \mathrm{m}, \quad \text{or} \quad \mathrm{m}/\text{Å} = 10^{10}$$

Substituting in the first equation gives the value of λ in ångström units

$$\lambda/\text{Å} = (\lambda/\mathrm{m})\ (\mathrm{m}/\text{Å}) = (5.896 \times 10^{-7})\ (10^{10}) = 5896$$

or

$$\lambda = 5896\ \text{Å}$$

Example 2. The vapour pressure of water at 20 °C is recorded to be

$$p(\mathrm{H_2O}, 20\ ^\circ\mathrm{C}) = 17.5\ \mathrm{Torr}$$

The torr, the bar, and the atmosphere are given by the equations (see table 7.2, under pressure)

$$\mathrm{Torr} \approx 133.3\ \mathrm{Pa},$$
$$\mathrm{bar} = 10^5\ \mathrm{Pa},$$
$$\mathrm{atm} = 101\,325\ \mathrm{Pa}.$$

Thus

$$\begin{aligned} p(\mathrm{H_2O}, 20\ ^\circ\mathrm{C}) &= 17.5 \times 133.3\ \mathrm{Pa} = 2.33\ \mathrm{kPa} \\ &= (2.33 \times 10^3/10^5)\ \mathrm{bar} = 23.3\ \mathrm{mbar} \\ &= (2.33 \times 10^3/101\,325)\ \mathrm{atm} = 2.30 \times 10^{-2}\ \mathrm{atm} \end{aligned}$$

Example 3. Spectroscopic measurements show that for the methylene radical, CH_2, the $\tilde{a}\ ^1A_1$ excited state lies at a wavenumber 3156 $\mathrm{cm^{-1}}$ above the $\tilde{X}\ ^3B_1$ ground state

$$\tilde{\nu}(\tilde{a} - \tilde{X}) = T_0(\tilde{a}) - T_0(\tilde{X}) = 3156\ \mathrm{cm^{-1}}$$

The excitation energy from the ground triplet state to the excited singlet state is thus

$$\begin{aligned} \Delta E = hc\tilde{\nu} &= (6.626 \times 10^{-34}\ \mathrm{J\,s})\ (2.998 \times 10^8\ \mathrm{m\,s^{-1}})\ (3156\ \mathrm{cm^{-1}}) \\ &= 6.269 \times 10^{-22}\ \mathrm{J\,m\,cm^{-1}} \\ &= 6.269 \times 10^{-20}\ \mathrm{J} = 6.269 \times 10^{-2}\ \mathrm{aJ} \end{aligned}$$

where the values of h and c are taken from the fundamental physical constants in chapter 5, and we

(1) A more appropriate name for 'quantity calculus' might be 'algebra of quantities', because it is the principles of algebra rather than calculus that are involved.

have used the relation m = 100 cm, or m $\mathrm{cm^{-1}}$ = 100. Since the electronvolt is given by the equation (table 7.2, under energy) eV ≈ 1.6022×10^{-19} J, or aJ ≈ (1/0.16022) eV

$$\Delta E = (6.269 \times 10^{-2}/0.16022)\ \mathrm{eV} = 0.3913\ \mathrm{eV}$$

Similarly the Hartree energy is given by (table 7.3) $E_\mathrm{h} = \hbar^2/m_\mathrm{e}a_0{}^2 \approx 4.3598$ aJ, or aJ ≈ $(1/4.3598)E_\mathrm{h}$, and thus the excitation energy is given in atomic units by

$$\Delta E = (6.269 \times 10^{-2}/4.3598)E_\mathrm{h} = 1.4380 \times 10^{-2}\ E_\mathrm{h}$$

Finally the molar excitation energy is given by

$$\begin{aligned}\Delta E_\mathrm{m} &= L\Delta E \\ &= (6.022 \times 10^{23}\ \mathrm{mol^{-1}})\ (6.269 \times 10^{-2}\ \mathrm{aJ}) \\ &= 37.75\ \mathrm{kJ\ mol^{-1}}\end{aligned}$$

Also, since kcal = 4.184 kJ, or kJ = (1/4.184) kcal,

$$\Delta E_\mathrm{m} = (37.75/4.184)\ \mathrm{kcal\ mol^{-1}} = 9.023\ \mathrm{kcal\ mol^{-1}}$$

Note that in this example the conversion factors are not pure numbers, but have dimensions, and involve the fundamental physical constants h, c, e, m_e, a_0 and L. Also in this example the necessary conversion factors could have been taken directly from the table on the inside back cover.

Example 4. The molar conductivity, Λ, of an electrolyte is defined by the equation (see p.60)

$$\Lambda = \kappa/c$$

where κ is the conductivity of the electrolyte solution minus the conductivity of the pure solvent and c is the electrolyte concentration. Conductivities of electrolytes are usually expressed in S $\mathrm{cm^{-1}}$ and concentrations in mol $\mathrm{dm^{-3}}$; for example, $\kappa(\mathrm{KCl}) = 7.39 \times 10^{-5}\ \mathrm{S\ cm^{-1}}$ for $c(\mathrm{KCl}) = 0.000\,500\ \mathrm{mol\ dm^{-3}}$. The molar conductivity can then be calculated as follows

$$\begin{aligned}\Lambda &= (7.39 \times 10^{-5}\ \mathrm{S\ cm^{-1}})/(0.000\,500\ \mathrm{mol\ dm^{-3}}) \\ &= 0.1478\ \mathrm{S\ mol^{-1}\ cm^{-1}\ dm^3} = 147.8\ \mathrm{S\ mol^{-1}\ cm^2}\end{aligned}$$

since $\mathrm{dm^3} = 1000\ \mathrm{cm^3}$. The above relationship has previously often been, and sometimes still is, written in the form

$$\Lambda = 1000\kappa/c$$

However, in this form the symbols *do not* represent physical quantities, but the *numerical values* of physical quantities in certain units. Specifically, the last equation is true only if Λ is the molar conductivity in S $\mathrm{mol^{-1}\ cm^2}$, κ is the conductivity in S $\mathrm{cm^{-1}}$, and c is the concentration in mol $\mathrm{dm^{-3}}$. This form does not follow the rules of quantity calculus, and should be avoided. The equation $\Lambda = \kappa/c$, in which the symbols represent physical quantities, is true in any units. If it is desired to write the relationship between numerical values it should be written in the form

$$\Lambda/(\mathrm{S\ mol^{-1}\ cm^2}) = \frac{1000\kappa/(\mathrm{S\ cm^{-1}})}{c/(\mathrm{mol\ dm^{-3}})}$$

Example 5. A solution of 0.125 mol of solute B in 953 g of solvent S has a molality m_B given by[2]

$$m_\mathrm{B} = n_\mathrm{B}/m_\mathrm{S} = (0.125/953)\ \mathrm{mol\ g^{-1}} = 0.131\ \mathrm{mol\ kg^{-1}}$$

(2) Note the confusion of notation: m_B denotes molality, and m_S denotes mass. However, these symbols are almost always used. See footnote (16) p.42.

The mole fraction of solute is approximately given by

$$x_B = n_B/(n_S + n_B) \approx n_B/n_S = m_B M_S$$

where it is assumed that $n_B \ll n_S$.

If the solvent is water with molar mass 18.015 g mol^{-1}, then

$$x_B \approx (0.131 \text{ mol kg}^{-1})\,(18.015 \text{ g mol}^{-1}) = 2.36 \text{ g/kg} = 0.00236$$

The equations used here are sometimes quoted in the form $m_B = 1000 n_B/m_S$, and $x_B \approx m_B M_S/1000$. However, this is *not* a correct use of quantity calculus because in this form the symbols denote the *numerical values* of the physical quantities in particular units; specifically it is assumed that m_B, m_S and M_S denote numerical values in mol kg^{-1}, g, and g mol^{-1} respectively. A correct way of writing the second equation would, for example, be

$$x_B = (m_B/\text{mol kg}^{-1})\,(M_S/\text{g mol}^{-1})/1000$$

Example 6. For paramagnetic materials the magnetic susceptibility may be measured experimentally and used to give information on the molecular magnetic dipole moment, and hence on the electronic structure of the molecules in the material. The paramagnetic contribution to the molar magnetic susceptibility of a material, χ_m, is related to the molecular magnetic dipole moment m by the Curie relation

$$\chi_m = \chi V_m = \mu_0 N_A m^2/3kT$$

In terms of the irrational susceptibility $\chi^{(ir)}$, which is often used in connection with the older esu, emu, and Gaussian unit systems (see section 7.3 below), this equation becomes

$$\chi_m^{(ir)} = \chi^{(ir)} V_m = (\mu_0/4\pi)\, N_A m^2/3kT$$

Solving for m, and expressing the result in terms of the Bohr magneton μ_B,

$$m/\mu_B = (3k/\mu_0 N_A)^{1/2} \mu_B^{-1} (\chi_m T)^{1/2}$$

Finally, using the values of the fundamental constants μ_B, k, μ_0, and N_A given in chapter 5, we obtain

$$\begin{aligned} m/\mu_B &= 0.7977[\chi_m/(\text{cm}^3 \text{ mol}^{-1})]^{1/2} [T/\text{K}]^{1/2} \\ &= 2.828[\chi_m^{(ir)}/(\text{cm}^3 \text{ mol}^{-1})]^{1/2} [T/\text{K}]^{1/2}. \end{aligned}$$

These expressions are convenient for practical calculations. The final result has frequently been expressed in the form

$$m/\mu_B = 2.828\,(\chi_m T)^{1/2}$$

where it is assumed, contrary to the conventions of quantity calculus, that χ_m and T denote the *numerical values* of the molar susceptibility and the temperature in the units cm^3 mol^{-1} and K respectively, and where it is also assumed (but rarely stated) that the susceptibility is defined using the irrational electromagnetic equations (see section 7.3 below).

7.2 CONVERSION TABLES FOR UNITS

The table below gives conversion factors from a variety of units to the corresponding SI unit. Examples of the use of this table have already been given in the preceding section. For each physical quantity the name is given, followed by the recommended symbol(s). Then the SI unit is given, followed by the esu, emu, Gaussian unit (Gau), atomic unit (au), and other units in common use, with their conversion factors to SI. The constant ζ which occurs in some of the electromagnetic conversion factors is the (exact) pure number $2.997\,924\,58 \times 10^{10} = c_0/(\text{cm s}^{-1})$.

The inclusion of non-SI units in this table should not be taken to imply that their use is to be encouraged. With some exceptions, SI units are always to be preferred to non-SI units. However, since many of the units below are to be found in the scientific literature, it is convenient to tabulate their relation to the SI.

For convenience units in the esu and Gaussian systems are quoted in terms of the four dimensions *length, mass, time*, and *electric charge*, by including the franklin (Fr) as an abbreviation for the electrostatic unit of charge and $4\pi\varepsilon_0$ as a constant with dimensions $(charge)^2/(energy \times length)$. This gives each physical quantity the same dimensions in all systems, so that all conversion factors are pure numbers. The factors $4\pi\varepsilon_0$ and the Fr may be eliminated by writing $\text{Fr} = \text{esu of charge} = \text{erg}^{1/2}\,\text{cm}^{1/2} = \text{cm}^{3/2}\,\text{g}^{1/2}\,\text{s}^{-1}$, and $4\pi\varepsilon_0 = \varepsilon_0^{(\text{ir})} = 1\ \text{Fr}^2\ \text{erg}^{-1}\,\text{cm}^{-1} = 1$, to recover esu expressions in terms of three base units (see section 7.3 below). The symbol Fr should be regarded as a compact representation of (esu of charge).

Conversion factors are either given exactly (when the = sign is used), or they are given to the approximation that the corresponding physical constants are known (when the ≈ sign is used). In the latter case the uncertainty is always less than ±5 in the last digit quoted.

Name	*Symbol*	*Relation to SI*
length, l		
metre (SI unit)	m	
centimetre (cgs unit)	cm	$= 10^{-2}$ m
bohr (au)	a_0, b	$= 4\pi\varepsilon_0\hbar^2/m_e e^2 \approx 5.291\,77 \times 10^{-11}$ m
ångström	Å	$= 10^{-10}$ m
micron	μ	$= \mu\text{m} = 10^{-6}$ m
millimicron	mμ	$= \text{nm} = 10^{-9}$ m
x unit	X	$\approx 1.002 \times 10^{-13}$ m
fermi	f, fm	$= \text{fm} = 10^{-15}$ m
inch	in	$= 2.54 \times 10^{-2}$ m
foot	ft	= 12 in = 0.3048 m
yard	yd	= 3 ft = 0.9144 m
mile	mi	= 1760 yd = 1609.344 m
nautical mile		= 1852 m
astronomical unit	AU	$= 1.496\,00 \times 10^{11}$ m
parsec	pc	$\approx 3.085\,68 \times 10^{16}$ m
light year	l.y.	$\approx 9.460\,528 \times 10^{15}$ m
light second		= 299 792 458 m
area, A		
square metre (SI unit)	m^2	
barn	b	$= 10^{-28}\ \text{m}^2$
acre		$\approx 4046.856\ \text{m}^2$
are	a	$= 100\ \text{m}^2$
hectare	ha	$= 10^4\ \text{m}^2$

Name	Symbol	Relation to SI
volume, V		
cubic metre (SI unit)	m^3	
litre	l, L	$= dm^3 = 10^{-3}\ m^3$
lambda	λ	$= \mu l = 10^{-6}\ dm^3$
barrel (US)		$\approx 158.987\ dm^3$
gallon (US)	gal (US)	$= 3.78541\ dm^3$
gallon (UK)	gal (UK)	$= 4.54609\ dm^3$
mass, m		
kilogram (SI unit)	kg	
gram (cgs unit)	g	$= 10^{-3}\ kg$
electron mass (au)	m_e	$\approx 9.10939 \times 10^{-31}\ kg$
unified atomic mass unit, dalton	u, Da	$= m_a(^{12}C)/12 \approx 1.660540 \times 10^{-27}\ kg$
gamma	γ	$= \mu g$
tonne	t	$= Mg = 10^3\ kg$
pound (avoirdupois)	lb	$= 0.45359237\ kg$
ounce (avoirdupois)	oz	$\approx 28.3495\ g$
ounce (troy)	oz (troy)	$\approx 31.1035\ g$
grain	gr	$= 64.79891\ mg$
time, t		
second (SI, cgs unit)	s	
au of time	$\hbar/E_h$	$\approx 2.41888 \times 10^{-17}\ s$
minute	min	$= 60\ s$
hour	h	$= 3600\ s$
day[1]	d	$= 86400\ s$
year[2]	a	$\approx 31556952\ s$
svedberg	Sv	$= 10^{-13}\ s$
acceleration, a		
SI unit	$m\ s^{-2}$	
standard acceleration of free fall	g_n	$= 9.80665\ m\ s^{-2}$
gal, galileo	Gal	$= 10^{-2}\ m\ s^{-2}$

(1) Note that the day is not exactly defined in terms of the second since so-called leap-seconds are added or subtracted from the day semiannually in order to keep the annual average occurrence of midnight at 24:00 on the clock.

(2) The year is not commensurable with the day and not a constant. Prior to 1967, when the atomic standard was introduced, the tropical year 1900 served as the basis for the definition of the second. For the epoch 1900.0. it amounted to 365.242 198 79 d $\approx$ 31 556 925.975 s and it decreases by 0.530 seconds per century. The calender years are exactly defined in terms of the day:

Julian year = 365.25 d
Gregorian year = 365.2425 d.

The definition in the table corresponds to the Gregorian year. This is an average based on a year of length 365 days, with leap years of 366 days; leap years are taken *either* when the year is divisible by 4 but is not divisible by 100, *or* when the year is divisible by 400. Whether the year 3200 should be a leap year is still open, but this does not have to be resolved until sometime in the middle of the 32nd century.

Name	*Symbol*	*Relation to SI*
force, F		
newton (SI unit)[3]	N	$= \text{kg m s}^{-2}$
dyne (cgs unit)	dyn	$= \text{g cm s}^{-2} = 10^{-5}\ \text{N}$
au of force	E_h/a_0	$\approx 8.238\,73 \times 10^{-8}\ \text{N}$
kilogram-force	kgf	$= 9.806\,65\ \text{N}$
energy, U		
joule (SI unit)	J	$= \text{kg m}^2\ \text{s}^{-2}$
erg (cgs unit)	erg	$= \text{g cm}^2\ \text{s}^{-2} = 10^{-7}\ \text{J}$
hartree (au)	E_h	$= \hbar^2/m_e a_0^2 \approx 4.359\,75 \times 10^{-18}\ \text{J}$
rydberg	Ry	$= E_h/2 \approx 2.179\,87 \times 10^{-18}\ \text{J}$
electronvolt	eV	$= e \times \text{V} \approx 1.602\,18 \times 10^{-19}\ \text{J}$
calorie, thermochemical	cal_{th}	$= 4.184\ \text{J}$
calorie, international	cal_{IT}	$= 4.1868\ \text{J}$
15 °C calorie	cal_{15}	$\approx 4.1855\ \text{J}$
litre atmosphere	l atm	$= 101.325\ \text{J}$
British thermal unit	Btu	$= 1055.06\ \text{J}$
pressure, p		
pascal (SI unit)	Pa	$= \text{N m}^{-2} = \text{kg m}^{-1}\ \text{s}^{-2}$
atmosphere	atm	$= 101\,325\ \text{Pa}$
bar	bar	$= 10^5\ \text{Pa}$
torr	Torr	$= (101\,325/760)\ \text{Pa} \approx 133.322\ \text{Pa}$
millimetre of mercury (conventional)	mmHg	$= 13.5951 \times 980.665 \times 10^{-2}\ \text{Pa} \approx 133.322\ \text{Pa}$
pounds per square inch	psi	$\approx 6.894\,757 \times 10^3\ \text{Pa}$
power, P		
watt (SI unit)	W	$= \text{kg m}^2\ \text{s}^{-3}$
horse power	hp	$= 745.7\ \text{W}$
action, L, J (angular momentum)		
SI unit	J s	$= \text{kg m}^2\ \text{s}^{-1}$
cgs unit	erg s	$= 10^{-7}\ \text{J s}$
au of action	$\hbar$	$= h/2\pi \approx 1.054\,57 \times 10^{-34}\ \text{J s}$
dynamic viscosity, η		
SI unit	Pa s	$= \text{kg m}^{-1}\ \text{s}^{-1}$
poise	P	$= 10^{-1}\ \text{Pa s}$
centipoise	cP	$= \text{mPa s}$
kinematic viscosity, ν		
SI unit	$\text{m}^2\ \text{s}^{-1}$	
stokes	St	$= 10^{-4}\ \text{m}^2\ \text{s}^{-1}$

(3) 1 N is approximately the force exerted by the earth upon an apple.

Name	*Symbol*	*Relation to SI*
thermodynamic temperature, T		
kelvin (SI unit)	K	
degree Rankine[4]	°R	$= (5/9)$ K
entropy, S		
heat capacity, C		
SI unit	$J\ K^{-1}$	
clausius	Cl	$= cal_{th}/K = 4.184\ J\ K^{-1}$
molar entropy, S_m		
molar heat capacity, C_m		
SI unit	$J\ K^{-1}\ mol^{-1}$	
entropy unit	e.u.	$= cal_{th}\ K^{-1}\ mol^{-1} = 4.184\ J\ K^{-1}\ mol^{-1}$
molar volume, V_m		
SI unit	$m^3\ mol^{-1}$	
amagat[5]	amagat	$= V_m$ of real gas at 1 atm and 273.15 K $\approx 22.4 \times 10^{-3}\ m^3\ mol^{-1}$
amount density, $1/V_m$		
SI unit	$mol\ m^{-3}$	
amagat[5]	amagat	$= 1/V_m$ for a real gas at 1 atm and 273.15 K $\approx 44.6\ mol\ m^{-3}$
plane angle, α		
radian (SI unit)	rad	
degree	°	$= \text{rad} \times 2\pi/360 \approx (1/57.295\,78)$ rad
minute	′	$=$ degree/60
second	″	$=$ degree/3600
grade	grad	$= \text{rad} \times 2\pi/400 \approx (1/63.661\,98)$ rad
radioactivity, A		
becquerel (SI unit)	Bq	$= s^{-1}$
curie	Ci	$= 3.7 \times 10^{10}$ Bq
absorbed dose of radiation[6]		
gray (SI unit)	Gy	$= J\ kg^{-1}$
rad	rad	$= 0.01$ Gy
dose equivalent		
sievert (SI unit)	Sv	$= J\ kg^{-1}$
rem	rem	≈ 0.01 Sv

(4) $T/°R = (9/5)T/K$. Also, Celsius temperature θ is related to thermodynamic temperature T by the equation

$$\theta/°C = T/K - 273.15$$

Similarly Fahrenheit temperature θ_F is related to Celsius temperature θ by the equation

$$\theta_F/°F = (9/5)(\theta/°C) + 32$$

(5) The name 'amagat' is unfortunately used as a unit for both molar volume and amount density. Its value is slightly different for different gases, reflecting the deviation from ideal behaviour for the gas being considered.
(6) The unit röntgen, employed to express exposure to X or γ radiations, is equal to: $R = 2.58 \times 10^{-4}\ C\ kg^{-1}$.

Name	Symbol	Relation to SI
electric current, I		
ampere (SI unit)	A	
esu, Gau	$(10/\zeta)$ A	$\approx 3.335\,64 \times 10^{-10}$ A
biot (emu)	Bi	$= 10$ A
au	$eE_h/\hbar$	$\approx 6.623\,62 \times 10^{-3}$ A
electric charge, Q		
coulomb (SI unit)	C	= A s
franklin (esu, Gau)	Fr	$= (10/\zeta)$ C $\approx 3.335\,64 \times 10^{-10}$ C
emu (abcoulomb)		= 10 C
proton charge (au)	e	$\approx 1.602\,18 \times 10^{-19}$ C $\approx 4.803\,21 \times 10^{-10}$ Fr
charge density, ρ		
SI unit	$\mathrm{C\,m^{-3}}$	
esu, Gau	$\mathrm{Fr\,cm^{-3}}$	$= 10^7 \zeta^{-1}\,\mathrm{C\,m^{-3}} \approx 3.335\,64 \times 10^{-4}\,\mathrm{C\,m^{-3}}$
au	ea_0^{-3}	$\approx 1.081\,20 \times 10^{-12}\,\mathrm{C\,m^{-3}}$
electric potential, V, φ		
volt (SI unit)	V	$= \mathrm{J\,C^{-1}} = \mathrm{J\,A^{-1}\,s^{-1}}$
esu, Gau	$\mathrm{erg\,Fr^{-1}}$	$= \mathrm{Fr\,cm^{-1}}/4\pi\varepsilon_0 = 299.792\,458$ V
'$\mathrm{cm^{-1}}$' (footnote 7)	$e\,\mathrm{cm^{-1}}/4\pi\varepsilon_0$	$\approx 1.439\,97 \times 10^{-7}$ V
au	$e/4\pi\varepsilon_0 a_0$	$= E_h/e \approx 27.2114$ V
mean international volt		= 1.000 34 V
US international volt		= 1.000 330 V
electric resistance, R		
ohm (SI unit)	Ω	$= \mathrm{V\,A^{-1}} = \mathrm{m^2\,kg\,s^{-3}\,A^{-2}}$
mean international ohm		= 1.000 49 Ω
US international ohm		= 1.000 495 Ω
electric field, E		
SI unit	$\mathrm{V\,m^{-1}}$	$= \mathrm{J\,C^{-1}\,m^{-1}}$
esu, Gau	$\mathrm{Fr\,cm^{-2}}/4\pi\varepsilon_0$	$= 2.997\,924\,58 \times 10^4\,\mathrm{V\,m^{-1}}$
'$\mathrm{cm^{-2}}$' (footnote 7)	$e\,\mathrm{cm^{-2}}/4\pi\varepsilon_0$	$\approx 1.439\,97 \times 10^{-5}\,\mathrm{V\,m^{-1}}$
au	$e/4\pi\varepsilon_0 a_0^2$	$\approx 5.142\,21 \times 10^{11}\,\mathrm{V\,m^{-1}}$
electric field gradient, $E'_{\alpha\beta}$, $q_{\alpha\beta}$		
SI unit	$\mathrm{V\,m^{-2}}$	$= \mathrm{J\,C^{-1}\,m^{-2}}$
esu, Gau	$\mathrm{Fr\,cm^{-3}}/4\pi\varepsilon_0$	$= 2.997\,924\,58 \times 10^6\,\mathrm{V\,m^{-2}}$
'$\mathrm{cm^{-3}}$' (footnote 7)	$e\,\mathrm{cm^{-3}}/4\pi\varepsilon_0$	$\approx 1.439\,97 \times 10^{-3}\,\mathrm{V\,m^{-2}}$
au	$e/4\pi\varepsilon_0 a_0^3$	$\approx 9.717\,36 \times 10^{21}\,\mathrm{V\,m^{-2}}$

(7) The units in quotation marks for electric potential through polarizability may be found in the literature, although they are strictly incorrect; they should be replaced in each case by the units given in the symbol column. Thus, for example, when a quadrupole moment is quoted in '$\mathrm{cm^2}$', the correct unit is $e\,\mathrm{cm^2}$; and when a polarizability is quoted in '$\mathrm{Å^3}$', the correct unit is $4\pi\varepsilon_0\,\mathrm{Å^3}$.

Name	*Symbol*	*Relation to SI*
electric dipole moment, p, μ		
SI unit	C m	
esu, Gau	Fr cm	$\approx 3.335\,64 \times 10^{-12}$ C m
debye	D	$= 10^{-18}$ Fr cm $\approx 3.335\,64 \times 10^{-30}$ C m
'cm', dipole length[7]	e cm	$\approx 1.602\,18 \times 10^{-21}$ C m
au	ea_0	$\approx 8.478\,36 \times 10^{-30}$ C m
electric quadrupole moment, $Q_{\alpha\beta}$, $\Theta_{\alpha\beta}$, eQ		
SI unit	C m^2	
esu, Gau	Fr cm^2	$\approx 3.335\,64 \times 10^{-14}$ C m^{-2}
'cm^2', quadrupole area[7]	e cm^2	$\approx 1.602\,18 \times 10^{-23}$ C m^2
au	ea_0^2	$\approx 4.486\,55 \times 10^{-40}$ C m^2
polarizability, α		
SI unit	J^{-1} C^2 m^2	= F m^2
esu, Gau, 'cm^3' polarizability volume[7]	$4\pi\varepsilon_0$ cm^3	$\approx 1.112\,65 \times 10^{-16}$ J^{-1} C^2 m^2
'$Å^3$' (footnote 7)	$4\pi\varepsilon_0$ $Å^3$	$\approx 1.112\,65 \times 10^{-40}$ J^{-1} C^2 m^2
au	$4\pi\varepsilon_0 a_0^3$	$\approx 1.648\,78 \times 10^{-41}$ J^{-1} C^2 m^2
electric displacement, D		
(volume) polarization, P		
SI unit	C m^{-2}	
esu, Gau	Fr cm^{-2}	$= (10^5/\zeta)$ C m^{-2} $\approx 3.335\,64 \times 10^{-6}$ C m^{-2}

(But note: the use of the esu or Gaussian unit for electric displacement usually implies that the irrational displacement is being quoted, $D^{(\mathrm{ir})} = 4\pi D$. See section 7.4.)

Name	*Symbol*	*Relation to SI*
magnetic flux density, B (magnetic field)		
tesla (SI unit)	T	= J A^{-1} m^{-2} = V s m^{-2} = Wb m^{-2}
gauss (emu, Gau)	G	$= 10^{-4}$ T
au	$\hbar/ea_0^2$	$\approx 2.350\,52 \times 10^5$ T
magnetic flux, Φ		
weber (SI unit)	Wb	= J A^{-1} = V s
maxwell (emu, Gau)	Mx	= G cm^{-2} $= 10^{-8}$ Wb
magnetic field, H		
(volume) magnetization, M		
SI unit	A m^{-1}	= C s^{-1} m^{-1}
oersted (emu, Gau)	Oe	$= 10^3$ A m^{-1}

(But note: in practice the oersted, Oe, is only used as a unit for $H^{(\mathrm{ir})} = 4\pi H$; thus when $H^{(\mathrm{ir})} = 1$ Oe, $H = (10^3/4\pi)$ A m^{-1}. See section 7.4.)

Name	*Symbol*	*Relation to SI*
magnetic dipole moment, m, μ		
SI unit	$\mathrm{A\,m^2}$	$= \mathrm{J\,T^{-1}}$
emu, Gau	$\mathrm{erg\,G^{-1}}$	$= 10\,\mathrm{A\,cm^2} = 10^{-3}\,\mathrm{J\,T^{-1}}$
Bohr magneton[8]	μ_B	$= e\hbar/2m_\mathrm{e} \approx 9.274\,02 \times 10^{-24}\,\mathrm{J\,T^{-1}}$
au	$e\hbar/m_\mathrm{e}$	$= 2\mu_\mathrm{B} \approx 1.854\,80 \times 10^{-23}\,\mathrm{J\,T^{-1}}$
nuclear magneton	μ_N	$= (m_\mathrm{e}/m_\mathrm{p})\,\mu_\mathrm{B} \approx 5.050\,79 \times 10^{-27}\,\mathrm{J\,T^{-1}}$
magnetizability, ξ		
SI unit	$\mathrm{J\,T^{-2}}$	$= \mathrm{C^2\,m^2\,kg^{-1}}$
au	$e^2 a_0{}^2/m_\mathrm{e}$	$\approx 7.891\,04 \times 10^{-29}\,\mathrm{J\,T^{-2}}$
magnetic susceptibility, χ, κ		
SI unit	1	
emu, Gau	1	

(But note: in practice susceptibilities quoted in the context of emu or Gaussian units are always values for $\chi^{(\mathrm{ir})} = \chi/4\pi$; thus when $\chi^{(\mathrm{ir})} = 10^{-6}$, $\chi = 4\pi \times 10^{-6}$. See section 7.3.)

molar magnetic susceptibility, χ_m		
SI unit	$\mathrm{m^3\,mol^{-1}}$	
emu, Gau	$\mathrm{cm^3\,mol^{-1}}$	$= 10^{-6}\,\mathrm{m^3\,mol^{-1}}$

(But note: in practice the units $\mathrm{cm^3\,mol^{-1}}$ usually imply that the irrational molar susceptibility is being quoted, $\chi_\mathrm{m}{}^{(\mathrm{ir})} = \chi_\mathrm{m}/4\pi$; thus, for example if $\chi_\mathrm{m}{}^{(\mathrm{ir})} = -15 \times 10^{-6}\,\mathrm{cm^3\,mol^{-1}}$, which is often written as '-15 cgs ppm', then $\chi_\mathrm{m} = -1.88 \times 10^{-10}\,\mathrm{m^3\,mol^{-1}}$. See section 7.3.)

(8) The Bohr magneton μ_B is sometimes denoted BM (or B.M.), but this is not recommended.

7.3 THE esu, emu, GAUSSIAN AND ATOMIC UNIT SYSTEMS

The SI equations of electromagnetic theory are usually used with physical quantities in SI units, in particular the four units m, kg, s, and A for length, mass, time and electric current. The basic equations for the electrostatic force between charges Q_1 and Q_2, and for the electromagnetic force between current elements $I_1\,\mathrm{d}\boldsymbol{l}_1$ and $I_2\,\mathrm{d}\boldsymbol{l}_2$, in vacuum, are written

$$\boldsymbol{F} = Q_1 Q_2 \boldsymbol{r}/4\pi\varepsilon_0 r^3 \tag{1a}$$
$$\boldsymbol{F} = (\mu_0/4\pi) I_1\,\mathrm{d}\boldsymbol{l}_1 \times (I_2\,\mathrm{d}\boldsymbol{l}_2 \times \boldsymbol{r})/r^3 \tag{1b}$$

The physical quantities ε_0 and μ_0, the permittivity and permeability of vacuum, respectively, have the values

$$\varepsilon_0 = (10^7/4\pi c_0{}^2)\ \mathrm{kg^{-1}\,m^{-1}\,C^2} \approx 8.854\,188 \times 10^{-12}\ \mathrm{C^2\,m^{-1}\,J^{-1}} \tag{2a}$$
$$\mu_0 = 4\pi \times 10^{-7}\ \mathrm{N\,A^{-2}} \approx 1.256\,637 \times 10^{-6}\ \mathrm{N\,A^{-2}} \tag{2b}$$

The value of μ_0 results from the definition of the ampere (section 3.2), which is such as to give μ_0 the value in (2b). The value of ε_0 then results from the relation

$$\varepsilon_0 \mu_0 = 1/c_0{}^2 \tag{3}$$

where c_0 is the speed of light in vacuum.

The numerical constant 4π is introduced into the definitions of ε_0 and μ_0 because of the spherical symmetry involved in equations (1); in this way we avoid its appearance in later equations relating to systems with rectangular symmetry. When factors of 4π are introduced in this way, as in the SI, the equations are described as 'rationalized'. The alternative 'unrationalized' or 'irrational' form of the electromagnetic equations is discussed below.

Other systems of units and equations in common use in electromagnetic theory, in addition to the SI, are the esu system, the emu system, the Gaussian system, and the system of atomic units. The conversion from SI to these other systems may be understood in the following steps.

First, all of the alternative systems involve equations written in the irrational form, in place of the rationalized form used in the SI. This involves changes of factors of 4π, and the redefinition of certain physical quantities. Second, a particular choice of units is made in each case to give either ε_0 or μ_0 a simple chosen value. Third, in the case of the esu, emu, and Gaussian systems (but not in the case of atomic units) the system of four base units (and four independent dimensions) is dropped in favour of only three base units (and independent dimensions) by an appropriate choice of the definition of charge or current in terms of length, mass and time. All these changes are considered in more detail below. Finally, because of the complications resulting from the alternative choice of rational or irrational relations, and the alternative ways of choosing the base dimensions, the equations of electromagnetic theory are different in the different systems. These changes are summarized in table 7.4 which gives the conversion of equations between the SI and the alternative systems.

(i) The change to irrational quantities and equations

Equations (1) can be written in the alternative four-quantity irrational form by defining new quantities $\varepsilon_0{}^{(\mathrm{ir})}$ and $\mu_0{}^{(\mathrm{ir})}$, so that (1a,b) become

$$\boldsymbol{F} = Q_1 Q_2 \boldsymbol{r}/\varepsilon_0{}^{(\mathrm{ir})} r^3 \tag{4a}$$
$$\boldsymbol{F} = \mu_0{}^{(\mathrm{ir})} I_1\,\mathrm{d}\boldsymbol{l}_1 \times (I_2\,\mathrm{d}\boldsymbol{l}_2 \times \boldsymbol{r})/r^3 \tag{4b}$$

The new quantities are related to ε_0 and μ_0 by the equations

$$\varepsilon_0{}^{(\mathrm{ir})} = 4\pi\varepsilon_0 \tag{5a}$$
$$\mu_0{}^{(\mathrm{ir})} = \mu_0/4\pi \tag{5b}$$

When the equations of electromagnetic theory are written in this alternative irrational form, six other new quantities are defined in addition to $\varepsilon_0^{(ir)}$ and $\mu_0^{(ir)}$; namely $\varepsilon^{(ir)}$, $\mu^{(ir)}$, $D^{(ir)}$, $H^{(ir)}$, $\chi_e^{(ir)}$ (the electric susceptibility), and $\chi^{(ir)}$ (the magnetic susceptibility). The definitions of other quantities remain unchanged. In each case we denote the new quantities by a superscript (ir) for irrational. The new quantities are defined in terms of the old quantities by the equations

$$\varepsilon^{(ir)} = 4\pi\varepsilon \tag{6a}$$

$$\mu^{(ir)} = \mu/4\pi \tag{6b}$$

$$D^{(ir)} = 4\pi D \tag{7a}$$

$$H^{(ir)} = 4\pi \mathrm{H} \tag{7b}$$

$$\chi_e^{(ir)} = \chi_e/4\pi \tag{8a}$$

$$\chi^{(ir)} = \chi/4\pi \tag{8b}$$

All of the equations of electromagnetic theory can now be transformed from the SI into the irrational form by using equations (5a, b), (6a, b), (7a, b) and (8a, b) to eliminate ε_0, μ_0, ε, μ, D, H, χ_e, and χ from the SI equations in favour of the corresponding irrational quantities distinguished by a superscript (ir).

The notation of a superscript (ir), used here to distinguish irrational quantities from their rational counterparts, where the definitions differ, is clumsy. However, in the published literature it is unfortunately customary to use exactly the same symbol for the quantities ε, μ, D, H, χ_e, and χ whichever definition (and corresponding set of equations) is in use. It is as though atomic and molecular physicists were to use the same symbol h for Planck's constant and Planck's constant/2π. Fortunately the different symbols h and $\hbar$ have been adopted in this case, and so we are able to write equations like $h = 2\pi\hbar$. Without some distinction in the notation, equations like (5), (6), (7) and (8) are impossible to write, and it is then difficult to discuss the relations between the rationalized SI equations and quantities and their irrational esu and emu equivalents. This is the reason for the rather cumbersome notation adopted here to distinguish quantities defined by different equations in the different systems.

(ii) The esu system

The esu system is based on irrational equations and quantities, and may be described either in terms of four base units and four independent dimensions or, as is more usual, in terms of three base units and three independent dimensions.

When four base units are used, they are taken to be the cm, g and s for length, mass and time, and the franklin[1] (symbol Fr) for the esu of charge, 1 Fr being chosen to be of such a magnitude that $\varepsilon_0^{(ir)} = 1\ \mathrm{Fr^2/erg\ cm}$. An equivalent definition of the franklin is that two charges of 1 Fr, 1 cm apart in a vacuum, repel each other with a force of one dyne. Other units are then derived from these four by the usual rules for constructing a coherent set of units from a set of base units.

The alternative and more usual form of the esu system is built on only three base units and three independent dimensions. This is achieved by defining the dimension of charge to be the same as that of $[(\text{energy}) \times (\text{length})]^{1/2}$, so that $1\ \mathrm{Fr^2} = 1\ \mathrm{erg\,cm}$. The Fr then disappears as a unit, and the constant $\varepsilon_0^{(ir)}$ is dimensionless, and equal to 1, so that it may be omitted from all equations. Thus

(1) The name 'franklin', symbol Fr, for the esu of charge was suggested by Guggenheim more than 40 years ago (*Nature*, **148** (1941) 751). Although it has not been widely adopted, this name and symbol are used here for convenience as a compact expression for the esu of charge. The name 'statcoulomb' has also been used for the esu of charge.

equation (4a) for the force between charges in vacuum, for example, becomes simply

$$\boldsymbol{F} = Q_1 Q_2 \boldsymbol{r}/r^3 \qquad (9)$$

This also means that the permittivity of a dielectric medium, $\varepsilon^{(\mathrm{ir})}$, is exactly the same as the relative permittivity or dielectric constant ε_r, so that only one of these quantities is required—which is usually simply called the permittivity, ε. Finally, since $\varepsilon_0{}^{(\mathrm{ir})} = 1$, equations (3) and (5) require that $\mu_0{}^{(\mathrm{ir})} = 1/c_0{}^2$.

To summarize, the transformation of equations from the four-quantity SI to the three-quantity esu system is achieved by making the substitutions $\varepsilon_0 = 1/4\pi$, $\mu_0 = 4\pi/c_0{}^2$, $\varepsilon = \varepsilon_r/4\pi$, $D = D^{(\mathrm{ir})}/4\pi$, and $\chi_e = 4\pi\chi_e{}^{(\mathrm{ir})}$.

(iii) The emu system

The emu system is also based on irrational equations and quantities, and may similarly be described in terms of either four or three base units.

When described in terms of four base units, they are taken as the cm, g, s, and the unit of electric current, which we call the (emu of current). This is chosen to be of such a magnitude that $\mu_0{}^{(\mathrm{ir})} = 1\ \mathrm{cm\,g\,s^{-2}}\ (\text{emu of current})^{-2}$. An equivalent definition of the emu of current is that the force between two parallel wires, 1 cm apart in a vacuum, each carrying 1 emu of current, is 2 dyn per cm of wire. Comparison with the definition of the ampere then shows that 1 (emu of current) = 10 A. Other units are derived from these four by the usual rules.[2]

In the more usual description of the emu system only three base units and three independent dimensions are used. The dimension (electric current) is defined to be the same as that of $(\text{force})^{1/2}$, so that 1 $(\text{emu of current})^2 = 1\ \mathrm{g\,cm\,s^{-2}} = 1$ dyn. The (emu of current) then disappears as a unit, and the constant $\mu_0{}^{(\mathrm{ir})}$ is dimensionless and equal to 1, so that it may be omitted from all equations. Thus equation (4b) for the force between current elements in vacuum, for example, becomes simply

$$\boldsymbol{F} = I_1\, \mathrm{d}\boldsymbol{l}_1 \times (I_2\, \mathrm{d}\boldsymbol{l}_2 \times \boldsymbol{r})/\mathrm{r}^3 \qquad (10)$$

The permeability of a magnetic medium $\mu^{(\mathrm{ir})}$ is identical to the relative permeability or magnetic constant μ_r, and is simply called the permeability. Finally $\varepsilon_0{}^{(\mathrm{ir})} = 1/c_0{}^2$ in the emu system.

To summarize, the transformation from the four-quantity SI to the three-quantity emu system is achieved by making the substitutions $\mu_0 = 4\pi$, $\varepsilon_0 = 1/4\pi c_0{}^2$, $\mu = 4\pi\mu_r$, $H = H^{(\mathrm{ir})}/4\pi$, and $\chi = 4\pi\chi^{(\mathrm{ir})}$.

(iv) The Gaussian system

The Gaussian system is a mixture of the esu system and the emu system, expressed in terms of three base units, esu being used for quantities in electrostatics and emu for electrodynamics. It is thus a hybrid system, and this gives rise to complications in both the equations and the units.

In the usual form of the Gaussian system, the following quantities are defined as in the esu system: charge Q, current I, electric field $\boldsymbol{E}$, electric displacement $\boldsymbol{D}^{(\mathrm{ir})}$, electric potential V, polarization $\boldsymbol{P}$, electric dipole moment $\boldsymbol{p}$, electric susceptibility $\chi_e{}^{(\mathrm{ir})}$, polarizability $\boldsymbol{\alpha}$, and capacitance C.

The following quantities are defined as in the emu system: magnetic flux density $\boldsymbol{B}$, magnetic flux Φ, magnetic potential $\boldsymbol{A}$, magnetic field $\boldsymbol{H}^{(\mathrm{ir})}$, magnetization $\boldsymbol{M}$, magnetic susceptibility $\chi^{(\mathrm{ir})}$, magnetic dipole moment $\boldsymbol{m}$, and magnetizability ξ. Neither $\varepsilon_0{}^{(\mathrm{ir})}$ nor $\mu_0{}^{(\mathrm{ir})}$ appear in the Gaussian equations, both being set equal to 1; the permittivity $\varepsilon^{(\mathrm{ir})} = \varepsilon_r$, and the permeability $\mu^{(\mathrm{ir})} = \mu_r$. However, the effect of equation (3) is that each physical quantity in the esu system differs in magnitude and dimensions from the corresponding emu quantity by some power of c_0. Thus the conversion of each SI equation of electromagnetic theory into the Gaussian form introduces factors of c_0, which are required to ensure internal consistency.

---

(2) The name biot, symbol Bi, has been used for the (emu of current).

The transformations of the more important equations between the Gaussian system and the SI are given in table 7.4 below.

(v) Atomic units [8] (see also section 3.8, p.76)

The so-called 'atomic units' are fundamental constants (and combinations of such constants) that arise in atomic and molecular electronic structure calculations, which are conveniently treated as though they were units. They may be regarded as a coherent system of units built on the four independent dimensions of length, mass, time, and electric charge. (The remaining dimensions used in the SI do not arise in electronic structure calculations.) However atomic units are more conveniently defined by taking a different choice for the four base dimensions, namely: mass, charge, action (angular momentum), and length. We choose the base unit of mass to be the electron rest mass m_e, the base unit of charge to be the elementary charge e, the base unit of action to be $\hbar = h/2\pi$ (where h is the Planck constant), and the base unit of length a_0 to be given by $a_0 = 4\pi\varepsilon_0\hbar^2/m_e e^2$. Taking these four units as base units, it follows that the unit of energy, called the hartree and denoted E_h, is given by $E_h = \hbar^2/m_e a_0^2$, and that $4\pi\varepsilon_0 = e^2/E_h a_0$. (The last relation is analogous to the relation in the four-quantity esu system where $4\pi\varepsilon_0 = \mathrm{Fr}^2/\mathrm{erg\ cm}$.)

The atomic unit of energy E_h the hartree, is (approximately) twice the ionisation energy of the hydrogen atom in its 1s ground state. The atomic unit of length a_0, the bohr, is approximately the distance of maximum radial density from the nucleus in the 1s orbital of a hydrogen atom. Clearly only four of the five units m_e, e, $\hbar$, E_h and a_0 are independent; useful ways of writing the interrelation are:

$$E_h = \hbar^2/m_e a_0^2 = e^2/4\pi\varepsilon_0 a_0 = m_e e^4/(4\pi\varepsilon_0)^2\hbar^2. \tag{11}$$

Conversion factors from atomic units to the SI are included in table 7.2 (p.110), and the five atomic units which have special names and symbols (described above), as well as a number of other atomic units, are also listed in table 3.8 (p.76).

The importance of atomic units lies in the fact that *ab initio* calculations in theoretical chemistry necessarily give results in atomic units (i.e. as multiples of m_e, e, $\hbar$, E_h and a_0). They are sometimes described as the 'natural units' of electronic calculations in theoretical chemistry. Indeed the results of such calculations can only be converted to other units (such as the SI) by using the current best estimates of the physical constants m_e, e, $\hbar$, etc., themselves expressed in SI units. It is thus appropriate for theoretical chemists to express their results in atomic units, and for the reader to convert to other units as and when necessary. This is also the reason why atomic units are written in italic (sloping) type rather than in the roman (upright) type usually used for units: the atomic units are physical quantities chosen from the fundamental physical constants of electronic structure calculations. There is, however, no authority from CGPM for designating these quantities as 'units', despite the fact that they are treated as units and called 'atomic units' by workers in the field.

Some authors who use atomic units use the customary symbols for physical quantities to represent the numerical values of quantities in the form (*physical quantity*)/(*atomic unit*), so that all quantities appear as pure numbers. Thus, for example, the Schrödinger equation for the hydrogen atom is written in SI in the form

$$-(\hbar^2/2m_e)\nabla_r^2\psi - (e^2/4\pi\varepsilon_0 r)\psi = E\psi \tag{12}$$

where ∇_r denotes derivatives with respect to r. After dividing throughout by E_h and making use of (11), this becomes

$$-\tfrac{1}{2}a_0^2\nabla_r^2\psi - (a_0/r)\psi = (E/E_h)\psi \tag{13}$$

If we now define $\rho = r/a_0$, and $E' = E/E_h$, so that ρ and E' are dimensionless numbers giving the numerical values of r and E in atomic units, then (13) can be written

$$-\tfrac{1}{2}\nabla_\rho^2\psi - (1/\rho)\psi = E'\psi \tag{14}$$

where ∇_ρ denotes derivatives with respect to ρ. Equation (14), in which each coefficient of ψ is dimensionless, is commonly described as being 'expressed in atomic units', and is the form usually adopted by theoretical chemists. Although the power of dimensional analysis is lost in this form, the symbolism has the advantage of simplicity. In using this form it is helpful to distinguish the dimensionless quantities which are here denoted ρ and E' from the customary physical quantities r and E themselves, but many authors make no distinction in either the symbol or the name.

Some authors also use the symbol 'au' (or 'a.u.') for every atomic unit, in place of the appropriate combination of the explicit symbols m_e, e, $\hbar$, E_h and a_0. This should be avoided. Appropriate combinations of m_e, e, $\hbar$, E_h and a_0 for the atomic units of various physical quantities are given in tables 3.8 (p.76) and 7.2 (p.110).

Examples $E = -0.345\ E_h$, not -0.345 atomic units
$r = 1.567\ a_0$, not 1.567 a.u. or 1.567 au

7.4 TRANSFORMATION OF EQUATIONS OF ELECTROMAGNETIC THEORY BETWEEN THE SI, THE FOUR-QUANTITY IRRATIONAL FORM AND THE GAUSSIAN FORM

Note that the esu equations may be obtained from the four-quantity irrational equations by putting $\varepsilon_0^{(\text{ir})} = 1$, and $\mu_0^{(\text{ir})} = 1/c_0^2$; the emu equations may be obtained by putting $\mu_0^{(\text{ir})} = 1$, and $\varepsilon_0^{(\text{ir})} = 1/c_0^2$.

SI relation	*Four-quantity irrational relation*	*Gaussian relation*
force on a moving charge Q with velocity $\boldsymbol{v}$:		
$\boldsymbol{F} = Q(\boldsymbol{E} + \boldsymbol{v} \times \boldsymbol{B})$	$\boldsymbol{F} = Q(\boldsymbol{E} + \boldsymbol{v} \times \boldsymbol{B})$	$\boldsymbol{F} = Q(\boldsymbol{E} + \boldsymbol{v} \times \boldsymbol{B}/c_0)$
force between charges in vacuum:		
$\boldsymbol{F} = Q_1 Q_2 \boldsymbol{r}/4\pi\varepsilon_0 r^3$	$\boldsymbol{F} = Q_1 Q_2 \boldsymbol{r}/\varepsilon_0^{(\text{ir})} r^3$	$\boldsymbol{F} = Q_1 Q_2 \boldsymbol{r}/r^3$
potential around a charge in vacuum:		
$V = Q/4\pi\varepsilon_0 r$	$V = Q/\varepsilon_0^{(\text{ir})} r$	$V = Q/r$
relation between field and potential:		
$\boldsymbol{E} = -\text{grad}\, V$	$\boldsymbol{E} = -\text{grad}\, V$	$\boldsymbol{E} = -\text{grad}\, V$
field due to a charge distribution in vacuum:		
$\text{div}\, \boldsymbol{E} = \rho/\varepsilon_0$	$\text{div}\, \boldsymbol{E} = 4\pi\rho/\varepsilon_0^{(\text{ir})}$	$\text{div}\, \boldsymbol{E} = 4\pi\rho$
capacitance of a parallel plate condenser, area A, separation d:		
$C = \varepsilon_0 \varepsilon_r A/d$	$C = \varepsilon_0^{(\text{ir})} \varepsilon_r A/4\pi d$	$C = \varepsilon_r A/4\pi d$
electric dipole moment of a charge distribution:		
$\boldsymbol{p} = \int \rho \boldsymbol{r}\, \text{d}V$	$\boldsymbol{p} = \int \rho \boldsymbol{r}\, \text{d}V$	$\boldsymbol{p} = \int \rho \boldsymbol{r}\, \text{d}V$
potential around a dipole in vacuum:		
$V = \boldsymbol{p} \cdot \boldsymbol{r}/4\pi\varepsilon_0 r^3$	$V = \boldsymbol{p} \cdot \boldsymbol{r}/\varepsilon_0^{(\text{ir})} r^3$	$V = \boldsymbol{p} \cdot \boldsymbol{r}/r^3$
energy of a charge distribution in an electric field:		
$E_p = QV - \boldsymbol{p} \cdot \boldsymbol{E} + \cdots$	$E_p = QV - \boldsymbol{p} \cdot \boldsymbol{E} + \cdots$	$E_p = QV - \boldsymbol{p} \cdot \boldsymbol{E} + \cdots$
electric dipole moment induced by a field:		
$\boldsymbol{p} = \alpha \boldsymbol{E} + \cdots$	$\boldsymbol{p} = \alpha \boldsymbol{E} + \cdots$	$\boldsymbol{p} = \alpha \boldsymbol{E} + \cdots$
relations between $\boldsymbol{E}$, $\boldsymbol{D}$ and $\boldsymbol{P}$:		
$\boldsymbol{E} = (\boldsymbol{D} - \boldsymbol{P})/\varepsilon_0$	$\boldsymbol{E} = (\boldsymbol{D}^{(\text{ir})} - 4\pi \boldsymbol{P})/\varepsilon_0^{(\text{ir})}$	$\boldsymbol{E} = \boldsymbol{D}^{(\text{ir})} - 4\pi \boldsymbol{P}$
$\boldsymbol{E} = \boldsymbol{D}/\varepsilon_0 \varepsilon_r$	$\boldsymbol{E} = \boldsymbol{D}^{(\text{ir})}/\varepsilon_0^{(\text{ir})} \varepsilon_r$	$\boldsymbol{E} = \boldsymbol{D}^{(\text{ir})}/\varepsilon_r$
relations involving the electric susceptibility:		
$\varepsilon_r = 1 + \chi_e$	$\varepsilon_r = 1 + 4\pi \chi_e^{(\text{ir})}$	$\varepsilon_r = 1 + 4\pi \chi_e^{(\text{ir})}$
$\boldsymbol{P} = \chi_e \varepsilon_0 \boldsymbol{E}$	$\boldsymbol{P} = \chi_e^{(\text{ir})} \varepsilon_0^{(\text{ir})} \boldsymbol{E}$	$\boldsymbol{P} = \chi_e^{(\text{ir})} \boldsymbol{E}$
force between current elements in vacuum:		
$\boldsymbol{F} = \dfrac{\mu_0}{4\pi} \dfrac{I\, \text{d}\boldsymbol{l}_1 \times (I\, \text{d}\boldsymbol{l}_2 \times \boldsymbol{r})}{r^3}$	$\boldsymbol{F} = \dfrac{\mu_0^{(\text{ir})} I\, \text{d}\boldsymbol{l}_1 \times (I\, \text{d}\boldsymbol{l}_2 \times \boldsymbol{r})}{r^3}$	$\boldsymbol{F} = \dfrac{I\, d\boldsymbol{l}_1 \times (I\, \text{d}\boldsymbol{l}_2 \times \boldsymbol{r})}{c_0^2 r^3}$
force on a current element in a field:		
$\boldsymbol{F} = I\, \text{d}\boldsymbol{l} \times \boldsymbol{B}$	$\boldsymbol{F} = I\, \text{d}\boldsymbol{l} \times \boldsymbol{B}$	$\boldsymbol{F} = I\, \text{d}\boldsymbol{l} \times \boldsymbol{B}/c_0$

SI relation	*Four-quantity irrational relation*	*Gaussian relation*
potential due to a current element in vacuum:		
$\boldsymbol{A} = (\mu_0/4\pi)\,(I\,\mathrm{d}\boldsymbol{l}/r)$	$\boldsymbol{A} = \mu_0^{(\mathrm{ir})} I\,\mathrm{d}\boldsymbol{l}/r$	$\boldsymbol{A} = I\,\mathrm{d}\boldsymbol{l}/c_0 r$
relation between field and potential:		
$\boldsymbol{B} = \mathrm{curl}\,\boldsymbol{A}$	$\boldsymbol{B} = \mathrm{curl}\,\boldsymbol{A}$	$\boldsymbol{B} = \mathrm{curl}\,\boldsymbol{A}$
field due to a current element in vacuum:		
$\boldsymbol{B} = (\mu_0/4\pi)(I\,\mathrm{d}\boldsymbol{l}\times\boldsymbol{r}/r^3)$	$\boldsymbol{B} = \mu_0^{(\mathrm{ir})}\, I\,\mathrm{d}\boldsymbol{l}\times\boldsymbol{r}/r^3$	$\boldsymbol{B} = I\,\mathrm{d}\boldsymbol{l}\times\boldsymbol{r}/c_0 r^3$
field due to a current density $\boldsymbol{j}$ in vacuum:		
$\mathrm{curl}\,\boldsymbol{B} = \mu_0 \boldsymbol{j}$	$\mathrm{curl}\,\boldsymbol{B} = 4\pi\mu_0^{(\mathrm{ir})}\boldsymbol{j}$	$\mathrm{curl}\,\boldsymbol{B} = 4\pi\boldsymbol{j}/c_0$
magnetic dipole of a current loop of area $\mathrm{d}\boldsymbol{A}$:		
$\boldsymbol{m} = I\,\mathrm{d}\boldsymbol{A}$	$\boldsymbol{m} = I\,\mathrm{d}\boldsymbol{A}$	$\boldsymbol{m} = I\,\mathrm{d}\boldsymbol{A}/c_0$
potential around a magnetic dipole in vacuum:		
$\boldsymbol{A} = (\mu_0/4\pi)\,(\boldsymbol{m}\times\boldsymbol{r}/r^3)$	$\boldsymbol{A} = \mu_0^{(\mathrm{ir})}\,\boldsymbol{m}\times\boldsymbol{r}/r^3$	$\boldsymbol{A} = \boldsymbol{m}\times\boldsymbol{r}/c_0\, r^3$
energy of a magnetic dipole in a field:		
$E_\mathrm{p} = -\boldsymbol{m}\cdot\boldsymbol{B}$	$E_\mathrm{p} = -\boldsymbol{m}\cdot\boldsymbol{B}$	$E_\mathrm{p} = -\boldsymbol{m}\cdot\boldsymbol{B}$
magnetic dipole induced by a field:		
$\boldsymbol{m} = \xi\boldsymbol{B}$	$\boldsymbol{m} = \xi\boldsymbol{B}$	$\boldsymbol{m} = \xi\boldsymbol{B}$
relations between $\boldsymbol{B}$, $\boldsymbol{H}$ and $\boldsymbol{M}$:		
$\boldsymbol{B} = \mu_0(\boldsymbol{H} + \boldsymbol{M})$	$\boldsymbol{B} = \mu_0^{(\mathrm{ir})}\,(\boldsymbol{H}^{(\mathrm{ir})} + 4\pi\boldsymbol{M})$	$\boldsymbol{B} = \boldsymbol{H}^{(\mathrm{ir})} + 4\pi\boldsymbol{M}$
$\boldsymbol{B} = \mu_0\mu_\mathrm{r}\boldsymbol{H}$	$\boldsymbol{B} = \mu_0^{(\mathrm{ir})}\,\mu_\mathrm{r}\boldsymbol{H}^{(\mathrm{ir})}$	$\boldsymbol{B} = \mu_\mathrm{r}\,\boldsymbol{H}^{(\mathrm{ir})}$
relations involving the magnetic susceptibility:		
$\mu_\mathrm{r} = 1 + \chi$	$\mu_\mathrm{r} = 1 + 4\pi\chi^{(\mathrm{ir})}$	$\mu_\mathrm{r} = 1 + 4\pi\chi^{(\mathrm{ir})}$
$\boldsymbol{M} = \chi\boldsymbol{B}/\mu_0$	$\boldsymbol{M} = \chi^{(\mathrm{ir})}\boldsymbol{B}/\mu_0^{(\mathrm{ir})}$	$\boldsymbol{M} = \chi^{(\mathrm{ir})}\boldsymbol{B}$
Curie relation:		
$\chi_\mathrm{m} = V_\mathrm{m}\chi$	$\chi_\mathrm{m}^{(\mathrm{ir})} = V_\mathrm{m}\chi^{(\mathrm{ir})}$	$\chi_\mathrm{m}^{(\mathrm{ir})} = V_\mathrm{m}\chi^{(\mathrm{ir})}$
$= L\mu_0 m^2/3kT$	$= L\mu_0^{(\mathrm{ir})} m^2/3kT$	$= Lm^2/3kT$
Maxwell equations:		
$\mathrm{div}\,\boldsymbol{D} = \rho$	$\mathrm{div}\,\boldsymbol{D}^{(\mathrm{ir})} = 4\pi\rho$	$\mathrm{div}\,\boldsymbol{D}^{(\mathrm{ir})} = 4\pi\rho$
$\mathrm{div}\,\boldsymbol{B} = 0$	$\mathrm{div}\,\boldsymbol{B} = 0$	$\mathrm{div}\,\boldsymbol{B} = 0$
$\mathrm{curl}\,\boldsymbol{E} + \partial\boldsymbol{B}/\partial t = 0$	$\mathrm{curl}\,\boldsymbol{E} + \partial\boldsymbol{B}/\partial t = 0$	$\mathrm{curl}\,\boldsymbol{E} + \dfrac{1}{c_0}\dfrac{\partial\boldsymbol{B}}{\partial t} = 0$
$\mathrm{curl}\,\boldsymbol{H} - \partial\boldsymbol{D}/\partial t = 0$	$\mathrm{curl}\,\boldsymbol{H}^{(\mathrm{ir})} - \partial\boldsymbol{D}^{(\mathrm{ir})}/\partial t = 0$	$\mathrm{curl}\,\boldsymbol{H}^{(\mathrm{ir})} - \dfrac{1}{c_0}\,\partial\boldsymbol{D}^{(\mathrm{ir})}/\partial t = 0$
energy density of radiation:		
$U/V = (\boldsymbol{E}\cdot\boldsymbol{D} + \boldsymbol{B}\cdot\boldsymbol{H})/2$	$U/V = \dfrac{\boldsymbol{E}\cdot\boldsymbol{D}^{(\mathrm{ir})} + \boldsymbol{B}\cdot\boldsymbol{H}^{(\mathrm{ir})}}{8\pi}$	$U/V = \dfrac{\boldsymbol{E}\cdot\boldsymbol{D}^{(\mathrm{ir})} + \boldsymbol{B}\cdot\boldsymbol{H}^{(\mathrm{ir})}}{8\pi}$
rate of energy flow (Poynting vector):		
$\boldsymbol{S} = \boldsymbol{E}\times\boldsymbol{H}$	$\boldsymbol{S} = \boldsymbol{E}\times\boldsymbol{H}^{(\mathrm{ir})}/4\pi$	$\boldsymbol{S} = c_0\,\boldsymbol{E}\times\boldsymbol{H}^{(\mathrm{ir})}/4\pi$

8
Abbreviations and acronyms

Abbreviations and acronyms (words formed from the initial letters of groups of words that are frequently repeated) should be used sparingly. Unless they are well established (e.g. NMR, IR) they should always be defined once in any paper, and they should generally be avoided in titles and abstracts. Abbreviations used to denote physical quantities should if possible be replaced by the recommended symbol for the quantity (e.g. E_i rather than IP for ionization energy, see. p.20; ρ rather than dens. for mass density, see p.12). For further recommendations concerning abbreviations see [46].

A list of frequently used abbreviations and acronyms is given here in order to help readers, but not necessarily to encourage their universal usage. In many cases an acronym can be found written in lower case letters and in capitals. In the list which follows only the most common usage is given. More extensive lists for different spectroscopic methods have been published by IUPAC [47, 48] and by Wendisch [75].

AA	atomic absorption
AAS	atomic absorption spectroscopy
ac	alternating current
ACM	adiabatic channel model
ACT	activated complex theory
A/D	analog-to-digital
ADC	analog-to-digital converter
AES	Auger electron spectroscopy
AIUPS	angle-integrated ultraviolet photoelectron spectroscopy
AM	amplitude modulated
amu	atomic mass unit (symbol: u) (see p.75)
AO	atomic orbital
APS	appearance potential spectroscopy
ARAES	angle-resolved Auger electron spectroscopy
AS	Auger spectroscopy
ATR	attenuated total (internal) reflection
AU	astronomical unit (see p. 110)
au	atomic unit (see section 7.3, p.120)
bcc	body centred cubic
BET	Brunauer–Emmett–Teller
BIS	bremsstrahlung isochromat spectroscopy
BM	Bohr magneton (symbol: μ_B, see p.116)
bp	boiling point
Btu	British thermal unit (see p.112)
CARS	coherent anti-Stokes Raman scattering
CAS	complete active space
CAS–SCF	complete active space – self consistent field
CAT	computer average of transients
CCA	coupled cluster approximation
ccp	cubic close packed
CD	circular dichroism
CEELS	characteristic electron energy loss spectroscopy
CELS	characteristic energy loss spectroscopy
CEPA	coupled electron pair approximation
cgs	centimetre-gram-second
CI	chemical ionization
CI	configuration interaction
CIDEP	chemically induced dynamic electron polarization
CIDNP	chemically induced dynamic nuclear polarization
CIMS	chemical ionization mass spectroscopy
CNDO	complete neglect of differential overlap
CSRS	coherent Stokes Raman scattering
CT	charge transfer
CVD	chemical vapour deposition
CW	continuous wave
D/A	digital-to-analog
DAPS	disappearance potential spectroscopy

dc	direct current
DLVO	Derjaguin–Landau–Verwey–Overbeek
DME	dropping mercury electrode
DRIFTS	diffuse reflectance infrared Fourier transform spectroscopy
DSC	differential scanning calorimeter
DTA	differential thermal analysis
E1	elimination unimolecular
E2	elimination bimolecular
EC	electron capture
ECD	electron capture detector
ED	electron diffraction
EDA	electron donor–acceptor [complex]
EELS	electron energy loss spectroscopy
EI	electron impact ionization
EIS	electron impact spectroscopy
EL	electroluminescence
ELDOR	electron–electron double resonance
ELEED	elastic low energy electron diffraction
emf	electromotive force
emu	electromagnetic unit (see section 7.3, p.119)
ENDOR	electron–nuclear double resonance
EPR	electron paramagnetic resonance
ESCA	electron spectroscopy for chemical applications (or analysis), see XPS
ESR	electron spin resonance
esu	electrostatic unit (see section 7.3, p.118)
ETS	electron transmission spectroscopy, electron tunnelling spectroscopy
eu	entropy unit (see p.113)
EXAFS	extended X-ray absorption fine structure
EXAPS	electron excited X-ray appearance potential spectroscopy
FAB(MS)	fast atom bombardment (mass spectroscopy)
fcc	face centred cubic
FD	field desorption
FEESP	field-emitted electron spin-polarization [spectroscopy]
FEM	field emission [electron] microscopy
FES	field emission spectroscopy
FFT	fast Fourier transform
FI	field ionization
FID	flame ionization detector
FID	free induction decay
FIM	field-ion microscopy
FIMS	field-ion mass spectroscopy
FIR	far-infrared
FM	frequency modulated
FPD	flame photometric detector
FSR	free spectral range (see p.31)
FT	Fourier transform
FTD	flame thermionic detector
FTIR	Fourier transform infrared
FWHM	full width at half maximum

GC	gas chromatography
glc	gas–liquid chromatography
GM	Geiger–Müller
GTO	Gaussian-type orbital (see p.19)
GVB	generalized valence bond
hcp	hexagonal close packed
HEED	high energy electron diffraction
HEELS	high energy electron energy loss spectroscopy
HF	Hartree–Fock (see p.17)
hfs	hyperfine structure (hyperfine splitting)
HMDE	hanging mercury drop electrode
HMO	Hückel molecular orbital (see p.17)
HOMO	highest occupied molecular orbital
HPLC	high-performance liquid chromatography
HREELS	high-resolution electron energy-loss spectroscopy
HTS	Hadamard transform spectroscopy
HWP	half-wave potential
IC	integrated circuit
ICR	ion cyclotron resonance
id	inner diameter
IEP	isoelectric point
IEPA	independent electron pair approximation
IETS	inelastic electron tunnelling spectroscopy
ILEED	inelastic low energy electron diffraction
INDO	incomplete neglect of differential overlap
INDOR	internuclear double resonance
INS	inelastic neutron scattering
I/O	input–output
IP	ionization potential (symbol: E_i, see p.20)
IPES	inverse photoelectron spectroscopy
IPTS	international practical temperature scale
IR	infrared
IS	ionization spectroscopy
ISS	ion scattering spectroscopy
L	ligand
LASER	light amplification by stimulated emission of radiation
LC	liquid chromatography
LCAO	linear combination of atomic orbitals
L-CCA	linear coupled-cluster approximation
LCMO	linear combination of molecular orbitals
LED	light-emitting diode
LEED	low-energy electron diffraction
LEELS	low energy electron loss spectroscopy
LEES	low-energy electron scattering
LET	linear energy transfer
LIDAR	light detection and ranging
LIF	laser induced fluorescence

LIS	laser isotope separation
LMR	laser magnetic resonance
LUMO	lowest unoccupied molecular orbital
M	central metal
MAR	magic-angle rotation
MAS	magic-angle spinning
MASER	microwave amplification by stimulated emission of radiation
MBE	molecular beam epitaxy
MBGF	many body Green's function
MBPT	many body perturbation theory
MC	Monte Carlo
MCA	multichannel analyser
MCD	magnetic circular dichroism
MCSCF	multiconfiguration self-consistent field
MD	molecular dynamics
MINDO	modified incomplete neglect of differential overlap
MIR	mid-infrared
MKSA	metre-kilogram-second-ampere
MM	molecular mechanics
MO	molecular orbital
MOCVD	metal organic chemical vapour deposition
MOMBE	metal organic molecular beam epitaxy
MORD	magnetic optical rotatory dispersion
MOS	metal oxide semiconductor
mp	melting point
MPI	multiphoton ionization
MPPT	Möller–Plesset perturbation theory
MP-SCF	Möller–Plesset self-consistent field
MRD	magnetic rotatory dispersion
MRI	magnetic resonance imaging
MS	mass spectroscopy
MW	microwave
MW	molecular weight (symbol: M_r, see p. 41)
NCE	normal calomel electrode
NEXAFS	near edge X-ray absorption fine structure
NIR	near-infrared
NMR	nuclear magnetic resonance
NOE	nuclear Overhauser effect
NQR	nuclear quadrupole resonance
NTP	normal temperature and pressure
od	outside diameter
ODMR	optically detected magnetic resonance
ORD	optical rotatory dispersion
PAS	photoacoustic spectroscopy
PC	paper chromatography
PD	see PED

PED	photoelectron diffraction
PES	photoelectron spectroscopy
PIES	Penning ionization electron spectroscopy, see PIS
PIPECO	photoion-photoelectron coincidence [spectroscopy]
PIS	Penning ionization (electron) spectroscopy
ppb	part per billion
pphm	part per hundred million
ppm	part per million
PPP	Pariser–Parr–Pople
PS	see PES
pzc	point of zero charge
QMS	quadrupole mass spectrometer
RADAR	radiowave detection and ranging
RAIRS	reflection/absorption infrared spectroscopy
RBS	Rutherford (ion) back scattering
RD	rotatory dispersion
RDE	rotating disc electrode
RDF	radial distribution function
REM	reflection electron microscopy
REMPI	resonance enhanced multiphoton ionization
RF	radio frequency
RHEED	reflection high-energy electron diffraction
RHF	restricted Hartree–Fock
RKR	Rydberg–Klein–Rees [potential]
rms	root mean square
RRK	Rice–Ramsperger–Kassel [theory]
RRKM	Rice–Ramsperger–Kassel–Marcus [theory]
RRS	resonance Raman spectroscopy
RS	Raman spectroscopy
RSPT	Rayleigh–Schrödinger perturbation theory
S	singlet
SCE	saturated calomel electrode
SCF	self-consistent field (see p.17)
SDCI	singly and doubly excited configuration interaction
S_E	substitution electrophilic
SEFT	spin-echo Fourier transform
SEM	scanning [reflection] electron microscopy
SEP	stimulated emission pumping
SERS	surface-enhanced Raman spectroscopy
SESCA	scanning electron spectroscopy for chemical applications
SEXAFS	surface extended X-ray absorption fine structure
SF	spontaneous fission
SHE	standard hydrogen electrode
SI	le système international d'unités
SIMS	secondary ion mass spectroscopy
S_N1	substitution nucleophilic unimolecular
S_N2	substitution nucleophilic bimolecular

S_Ni	substitution nucleophilic intramolecular
SOR	synchrotron orbital radiation
SRS	synchrotron radiation source
STEM	scanning transmission [electron] microscopy
STM	scanning tunnelling (electron) microscopy
STO	Slater-type orbital (see p. 19)
STP	standard temperature and pressure
T	triplet
TCC	thermal conductivity cell
TCD	thermal conductivity detector
TCF	time correlation function
TDMS	tandem quadrupole mass spectroscopy
TDS	thermal desorption spectroscopy
TEM	transmission electron microscopy
TG	thermogravimetry
TGA	thermogravimetric analysis
tlc	thin layer chromatography
TOF	time-of-flight [analysis]
TPD	temperature programmed desorption
TR^3	time-resolved resonance Raman scattering
TST	transition state theory
UHF	unrestricted Hartree–Fock
UHF	ultra high frequency
UHV	ultra high vacuum
UPES	ultraviolet photoelectron spectroscopy
UPS	ultraviolet photoelectron spectroscopy
UV	ultraviolet
VB	valence bond
VCD	vibrational circular dichroism
VEELS	vibrational electron energy-loss spectroscopy
VHF	very high frequency
VIS	visible
VLSI	very large scale integration
VPC	vapour-phase chromatography
VSEPR	valence shell electron pair repulsion
VUV	vacuum ultraviolet
X	halogen
XANES	X-ray absorption near-edge structure [spectroscopy]
XAPS	X-ray appearance potential spectroscopy
XPD	X-ray photoelectron diffraction
XPES	X-ray photoelectron spectroscopy
XPS	X-ray photoelectron spectroscopy
XRD	X-ray diffraction
Y–AG	yttrium aluminium garnet
ZPE	zero point energy

9
References

9.1 PRIMARY SOURCES

1 Manual of Symbols and Terminology for Physicochemical Quantities and Units
 (a) 1st ed., McGlashan, M.L., *Pure Appl. Chem.* **21** (1970) 1–38.
 (b) 2nd ed., Paul, M.A., Butterworths, London 1975.
 (c) 3rd ed., Whiffen, D.H., *Pure Appl. Chem.* **51** (1979) 1–36.
 (d) Appendix I—Definitions of Activities and Related Quantities, Whiffen, D.H., *Pure Appl. Chem.* **51** (1979) 37–41.
 (e) Appendix II—Definitions, Terminology and Symbols in Colloid and Surface Chemistry, Part I, *Pure Appl. Chem.* **31** (1972) 577–638.
 (f) Section 1.13: Selected Definitions, Terminology and Symbols for Rheological Properties Lyklema, J. and van Olphen, H., *Pure Appl. Chem.* **51** (1979) 1213–1218.
 (g) Section 1.14: Light scattering, Kerker, M. and Kratohvil, J.P., *Pure Appl. Chem.* **55** (1983) 931–941.
 (h) Part II: Heterogeneous Catalysis, Burwell Jr., R.L., *Pure Appl. Chem.* **46** (1976) 71–90.
 (i) Appendix III—Electrochemical Nomenclature, *Pure Appl. Chem.* **37** (1974) 499–516.
 (j) Appendix IV—Notation for States and Processes, Significance of the Word "Standard" in Chemical Thermodynamics, and Remarks on Commonly Tabulated Forms of Thermodynamic Functions, Cox, J.D., *Pure Appl. Chem.* **54** (1982) 1239–1250.
 (k) Appendix V—Symbolism and Terminology in Chemical Kinetics, Jenkins, A.D., *Pure Appl. Chem.* **53** (1981) 753–771.

2 (a) Mills, I., Cvitaš, T., Homann, K., Kallay, N. and Kuchitsu, K., *Quantities, Units and Symbols in Physical Chemistry*, 1st edn. Blackwell Scientific Publications, Oxford 1988.
 (b) Nomenklaturniye pravila IUPAC po Khimii Vol. 6, Fizicheskaya Khimiya, Nacionalnii Komitet Sovetskih Khimikov, Moscow 1988.
 (c) Riedel, M., *A fizikai-kémiai definiciók és jelölések*, Tankönyvkiadó, Budapest 1990.
 (d) Kuchitsu, K., *Quantities, Units and Symbols in Physical Chemistry*, Kodansha, Tokyo 1991.

3 Bureau International des Poids et Mesures, *Le Système International d'Unités (SI)*, 6th French and English Edition, BIPM, Sèvres 1991.

4 Cohen, E.R. and Giacomo, P., *Symbols, Units, Nomenclature and Fundamental Constants in Physics*, 1987 Revision, Document I.U.P.A.P.-25 (IUPAP–SUNAMCO 87–1) also published in: *Physica* **146A** (1987) 1–68.

5 International Standards ISO
International Organization for Standardization, Geneva
 (a) ISO 31-0: 1992, Quantities and Units–Part 0: General Principles Units and Symbols
 (b) ISO 31-1: 1992, Quantities and Units–Part 1: Space and Time
 (c) ISO 31-2: 1992, Quantities and Units–Part 2: Periodic and Related Phenomena
 (d) ISO 31-3: 1992, Quantities and Units–Part 3: Mechanics
 (e) ISO 31-4: 1992, Quantities and Units–Part 4: Heat
 (f) ISO 31-5: 1992, Quantities and Units–Part 5: Electricity and Magnetism
 (g) ISO 31-6: 1992, Quantities and Units–Part 6: Light and Related Electromagnetic Radiations
 (h) ISO 31-7: 1992, Quantities and Units–Part 7: Acoustics
 (i) ISO 31-8: 1992, Quantities and Units–Part 8: Physical Chemistry and Molecular Physics
 (j) ISO 31-9: 1992, Quantities and Units–Part 9: Atomic and Nuclear Physics
 (k) ISO 31-10: 1992, Quantities and Units–Part 10: Nuclear Reactions and Ionizing Radiations
 (m) ISO 31-11: 1992, Quantities and Units–Part 11: Mathematical Signs and Symbols for Use in the Physical Sciences and Technology

(n) ISO 31-12: 1992, Quantities and Units–Part 12: Characteristic Numbers
(p) ISO 31-13: 1992, Quantities and Units–Part 13: Solid State Physics
6 ISO1000: 1992, SI Units and Recommendations for the Use of Their Multiples and of Certain Other Units

All the standards listed here (5–6) are jointly reproduced in the ISO Standards Handbook 2, *Quantities and Units*, ISO, Geneva 1993.

7 ISO2955-1983, Information Processing—Representations of SI and Other Units for Use in Systems with Limited Character Sets

9.2 IUPAC REFERENCES

8 Rigg, J.C., Visser, B.F. and Lehmann, H.P., Nomenclature of Derived Quantities, *Pure Appl. Chem.* **63** (1991) 1307–1311.

9 Whiffen, D.H., Expression of Results in Quantum Chemistry, *Pure Appl. Chem.* **50** (1978) 75–79.

10 Becker, E.D., Recommendations for Presentation of Infrared Absorption Spectra in Data Collections: A—Condensed Phases, *Pure Appl. Chem.* **50** (1978) 231–236.

11 Becker, E.D., Durig, J.R., Harris, W.C. and Rosasco, G.J., Presentation of Raman Spectra in Data Collections, *Pure Appl. Chem.* **53** (1981) 1879–1885.

12 Recommendations for the Presentation of NMR Data for Publication in Chemical Journals, *Pure Appl. Chem.* **29** (1972) 625–628.

13 Presentation of NMR Data for Publication in Chemical Journals: B—Conventions Relating to Spectra from Nuclei other than Protons, *Pure Appl. Chem.* **45** (1976) 217–219.

14 Nomenclature and Spectral Presentation in Electron Spectroscopy Resulting from Excitation by Photons, *Pure Appl. Chem.* **45** (1976) 221–224.

15 Nomenclature and Conventions for Reporting Mössbauer Spectroscopic Data, *Pure Appl. Chem.* **45** (1976) 211–216.

16 Beynon, J.H., Recommendations for Symbolism and Nomenclature for Mass Spectroscopy, *Pure Appl. Chem.* **50** (1978) 65–73.

17 Morino, Y. and Shimanouchi, T., Definition and Symbolism of Molecular Force Constants, *Pure Appl. Chem.* **50** (1978) 1707–1713.

18 Lamola, A.A. and Wrighton, M.S., Recommended Standards for Reporting Photochemical Data, *Pure Appl. Chem.* **56** (1984) 939–944.

19 Sheppard, N., Willis, H.A. and Rigg, J.C., Names, Symbols, Definitions and Units of Quantities in Optical Spectroscopy, *Pure Appl. Chem.* **57** (1985) 105–120.

20 Fassel, V.A., Nomenclature, Symbols, Units and their Usage in Spectrochemical Analysis. I: General Atomic Emission Spectroscopy, *Pure Appl. Chem.* **30** (1972) 651–679.

21 Melmish, W.H., Nomenclature, Symbols, Units and their Usage in Spectrochemical Analysis. VI: Molecular Luminescence Spectroscopy, *Pure Appl. Chem.* **56** (1984) 231–245.

22 Leigh, G.J., *Nomenclature of Inorganic Chemistry*, Blackwell Scientific Publications, Oxford 1990.

23 Rigaudy, J. and Klesney, S.P., *Nomenclature of Organic Chemistry, Sections A, B, C, D, E, F and H*, Pergamon Press, Oxford 1979.

24 Ewing, M.B., Lilley, T.H., Olofsson, G.M., Rätzsch, M.T. and Somsen, G., Standard Quantities in Chemical Thermodynamics, *Pure Appl. Chem.* **65** (1993) in press.

25 Cali, J.P. and Marsh, K.N., An Annotated Bibliography on Accuracy in Measurement, *Pure Appl. Chem.* **55** (1983) 907–930.

26 Olofsson, G., Assignment and Presentation of Uncertainties of the Numerical Results of Thermodynamic Measurements, *Pure Appl. Chem.* **53** (1981) 1805–1825.

27 Cornish-Bowden, A., Glossary of Terms Used in Physical Organic Chemistry, *Pure Appl. Chem.* **55** (1983) 1281–1371.

28 Braslavsky, S.E. and Houk, K.N., Glossary of Terms Used in Photochemistry, *Pure Appl. Chem.* **60** (1988) 1055–1106.

29 Bard, A.J., Memming, R. and Miller, B., Terminology in Semiconductor Electrochemistry and Photoelectrochemical Energy Conversion, *Pure Appl. Chem.* **63** (1991) 569–596.

30 Heusler, K.E., Landolt, D. and Trasatti, S., Electrochemical Corrosion Nomenclature, *Pure Appl. Chem.* **61** (1989) 19–22.

31 Trasatti, S., The Absolute Electrode Potential: an Explanatory Note, *Pure Appl. Chem.* **58** (1986) 955–966.

32 Parsons, R., Electrode Reaction Orders, Transfer Coefficients and Rate Constants: Amplifi-

cation of Definitions and Recommendations for Publication of Parameters, *Pure Appl. Chem.* **52** (1980) 233–240.

33 Ibl, N., Nomenclature for Transport Phenomena in Electrolytic Systems, *Pure Appl. Chem.* **53** (1981) 1827–1840.

34 van Rysselberghe, P., Bericht der Kommission für elektrochemische Nomenklatur und Definitionen, *Z. Electrochem.* **58** (1954) 530–535.

35 Bard, A.J., Parsons, R. and Jordan, J., *Standard Potentials in Aqueous Solutions*, Marcel Dekker Inc., New York 1985.

36 Covington, A.K., Bates, R.G. and Durst, R.A., Definition of pH Scales, Standard Reference Values, Measurement of pH and Related Terminology, *Pure Appl. Chem.* **57** (1985) 531–542.

37 Sing, K.S.W., Everett, D.H., Haul, R.A.W., Moscou, L., Pierotti, R.A., Rouquérol, J. and Siemieniewska, T., Reporting Physisorption Data for Gas/Solid Systems *Pure Appl. Chem.* **57** (1985) 603–619.

38 Ter-Minassian-Saraga, L., Reporting Experimental Pressure–Area Data with Film Balances, *Pure Appl. Chem.* **57** (1985) 621–632.

39 Everett, D.H., Reporting Data on Adsorption from Solution at the Solid/Solution Interface, *Pure Appl. Chem.* **58** (1986) 967–984.

40 Haber, J., Manual on Catalyst Characterization, *Pure Appl. Chem.* **63** (1991) 1227–1246.

41 Metanomski, W.V., *Compendium of Macromolecular Nomenclature*, Blackwell Scientific Publications, Oxford 1991.

42 Holden, N.E., Atomic Weights of the Elements 1979, *Pure Appl. Chem.* **52** (1980) 2349–2384.

43 Peiser, H.S., Holden, N.E., de Bièvre, P., Barnes, I.L., Hagemann, R., de Laeter, J.R., Murphy, T.J., Roth, E., Shima, M. and Thode, H.G., Element by Element Review of Their Atomic Weights, *Pure Appl. Chem.* **56** (1984) 695–768.

44 Atomic Weights of the Elements 1991, *Pure Appl. Chem.* **64** (1992) 1519–1534.

45 Isotopic Compositions of the Elements 1989, *Pure Appl. Chem.* **63** (1991) 991–1002.

46 Lide, D., Use of Abbreviations in the Chemical Literature, *Pure Appl. Chem.* **52** (1980) 2229–2232.

47 Porter, H.Q. and Turner, D.W., A Descriptive Classification of the Electron Spectroscopies, *Pure Appl. Chem.* **59** (1987) 1343–1406.

48 Sheppard, N., English-Derived Abbreviations for Experimental Techniques in Surface Science and Chemical Spectroscopy, *Pure Appl. Chem.* **63** (1991) 887–893.

9.3 ADDITIONAL REFERENCES

49 Mullay, J., Estimation of atomic and group electronegativities, *Structure and Bonding* **66** (1987) 1–25.

50 Jenkins, F.A., Notation for the Spectra of Diatomic Molecules, *J. Opt. Soc. Amer.* **43** (1953) 425–426.

51 Mulliken, R.S., Report on Notation for the Spectra of Polyatomic Molecules, *J. Chem. Phys.* **23** (1955) 1997–2011. (Erratum *J. Chem. Phys.* **24** (1956) 1118.)

52 Herzberg, G., *Molecular Spectra and Molecular Structure* Vol. I. *Spectra of Diatomic Molecules*, Van Nostrand, Princeton 1950. Vol. II. *Infrared and Raman Spectra of Polyatomic Molecules*, Van Nostrand, Princeton 1946. Vol. III. *Electronic Spectra and Electronic Structure of Polyatomic Molecules*, Van Nostrand, Princeton 1966.

53 Watson, J.K.G., Aspects of Quartic and Sextic Centrifugal Effects on Rotational Energy Levels. In: Durig, J. R. (ed), *Vibrational Spectra and Structure*, Vol. 6, Elsevier, Amsterdam 1977, pp 1–89.

54 (a) Callomon, J.H., Hirota, E., Kuchitsu, K., Lafferty, W.J., Maki, A.G. and Pote, C.S., Structure Data of Free Polyatomic Molecules. In: Hellwege, K.-H. and Hellwege, A.M. (eds), *Landolt–Börnstein*, New Series, II/7, Springer Verlag, Berlin 1976.

(b) Callomon, J. H., Hirota, E., Iijima, T., Kuchitsu, K. and Lafferty, W., Structure Data of Free Polyatomic Molecules. In: Hellwege, K.-H. and Hellwege, A.M. (eds), *Landolt–Börnstein*, New Series, II/15 (Supplement to II/7), Springer Verlag, Berlin 1987.

55 Bunker, P.R., *Molecular Symmetry and Spectroscopy*, Academic Press, New York 1979.

56 Brown, J.M., Hougen, J.T., Huber, K.-P., Johns, J.W.C., Kopp, I., Lefebvre-Brion, H., Merer, A.J., Ramsay, D.A., Rostas, J. and Zare, R.N., The Labeling of Parity Doublet Levels in Linear Molecules, *J. Mol. Spectrosc.* **55** (1975) 500–503.

57 Alexander, M.H., Andresen, P., Bacis, R., Bersohn, R., Comes, F.J., Dagdigian, P.J., Dixon, R.N., Field, R.W., Flynn, G.W., Gericke, K.-H., Grant, E.R., Howard, B.J., Huber, J.R., King, D.S., Kinsey, J.L., Kleinermanns, K., Kuchitsu, K., Luntz, A.C., McCaffery, A. J., Pouilly, B., Reisler, H., Rosenwaks, S., Rothe, E.W., Shapiro, M., Simons, J.P., Vasudev, R., Wiesenfeld, J.R., Wittig, C. and Zare, R.N., A Nomenclature for Λ-doublet Levels in Rotating Linear Molecules, *J. Chem. Phys.* **89** (1988) 1749–1753.

58 Brand, J.C.D., Callomon, J.H., Innes, K.K., Jortner, J., Leach, S., Levy, D.H., Merer, A.J., Mills, I.M., Moore, C.B., Parmenter, C.S., Ramsay, D.A., Narahari Rao, K., Schlag, E.W., Watson, J.K.G. and Zare, R.N., The Vibrational Numbering of Bands in the Spectra of Polyatomic Molecules, *J. Mol. Spectrosc.* **99** (1983) 482–483.

59 Quack, M., Spectra and Dynamics of Coupled Vibrations in Polyatomic Molecules, *Ann. Rev. Phys. Chem.* **41** (1990) 839–874.

60 Maki, A.G. and Wells, J.S., Wavenumber Calibration Tables from Heterodyne Frequency Measurements, *NIST Special Publication 821*, U.S. Department of Commerce, 1991.

61 (a) Pugh, L.A. and Rao, K.N., Intensities from Infrared Spectra. In: Rao, K.N. (ed), *Molecular Spectroscopy: Modern Research*, Vol. II, Academic Press, New York 1976, pp.165–227.

(b) Smith, M.A., Rinsland, C.P., Fridovich, B. and Rao, K.N., Intensities and Collision Broadening Parameters from Infrared Spectra. In: Rao, K.N. (ed), *Molecular Spectroscopy: Modern Research*, Vol. III, Academic Press, New York 1985, pp.111–248.

62 Hahn, Th. (ed), *International Tables for Crystallography*, Vol. A, 2nd edn: *Space–Group Symmetry*, Reidel Publishing Co., Dordrecht 1983.

63 Alberty, R.A., Chemical Equations are Actually Matrix Equations, *J. Chem. Educ.* **68** (1991) 984.

64 Domalski, E.S., Selected Values of Heats of Combustion and Heats of Formation of Organic Compounds, *J. Phys. Chem. Ref. Data* **1** (1972) 221–277.

65 Freeman, R.D., Conversion of Standard (1 atm) Thermodynamic Data to the New Standard State Pressure, 1 bar (10^5 Pa), *Bull. Chem. Thermodyn.* **25** (1982) 523–530, *J. Chem. Eng. Data* **29** (1984) 105–111, *J. Chem. Educ.* **62** (1985) 681–686.
66 Wagman, D.D., Evans, W.H., Parker, V.B., Schumm, R.H., Halow, I., Bailey, S.M., Churney, K.L. and Nuttall, R.L., The NBS Tables of Chemical Thermodynamic Properties, *J. Phys. Chem. Ref. Data* **11** Suppl. 2 (1982) 1–392.
67 Chase, M.W., Davies, C.A., Downey, J.R., Frurip, D.J., McDonald, R.A. and Syverud, A.N., JANAF Thermochemical Tables, 3rd edn. *J. Phys. Chem. Ref. Data* **14** Suppl. 1 (1985) 1–392.
68 Glushko, V.P. (ed), *Termodinamicheskie svoistva individualnykh veshchestv*, Vols. 1–4, Nauka, Moscow 1978–85.
69 CODATA Task Group on Data for Chemical Kinetics: The Presentation of Chemical Kinetics Data in the Primary Literature, *CODATA Bull.* **13** (1974) 1–7.
70 Cohen, E.R. and Taylor, B.N., The 1986 Adjustment of the Fundamental Physical Constants, *CODATA Bull.* **63** (1986) 1–49.
71 Particle Data Group, 1992 Review of Particle Properties, *Phys. Rev.* **D45**, Part 2 (1992).
72 Wapstra, A.H. and Audi, G., The 1983 Atomic Mass Evaluation. I. Atomic Mass Table, *Nucl. Phys.* **A432** (1985) 1–54.
73 Raghavan, P., Table of Nuclear Moments, *Atomic Data Nucl. Data Tab.* **42** (1989) 189–291.
74 Pyykkö, P., The Nuclear Quadrupole Moments of the First 20 Elements: High Precision Calculations on Atoms and Small Molecules, *Z. Naturforsch.* **A47** (1992) 189–196.
75 Wendisch, D.A.W., *Acronyms and Abbreviations in Molecular Spectroscopy*, Springer Verlag, Heidelberg 1990.

THE GREEK ALPHABET

Α, α	A, α	*Alpha*	Ν, ν	N, ν	*Nu*
Β, β	B, β	*Beta*	Ξ, ξ	Ξ, ξ	*Xi*
Γ, γ	Γ, γ	*Gamma*	Ο, ο	O, o	*Omicron*
Δ, δ	Δ, δ	*Delta*	Π, π	Π, π	*Pi*
Ε, ε	E, ε	*Epsilon*	Ρ, ρ	P, ρ	*Rho*
Ζ, ζ	Z, ζ	*Zeta*	Σ, σ	Σ, σ	*Sigma*
Η, η	H, η	*Eta*	Τ, τ	T, τ	*Tau*
Θ, ϑ, θ	$\Theta, \vartheta, \theta$	*Theta*	ϒ, υ	Υ, υ	*Upsilon*
Ι, ι	I, ι	*Iota*	Φ, φ, ϕ	Φ, φ, ϕ	*Phi*
Κ, ϰ	K, κ	*Kappa*	Χ, χ	X, χ	*Chi*
Λ, λ	Λ, λ	*Lambda*	Ψ, ψ	Ψ, ψ	*Psi*
Μ, μ	M, μ	*Mu*	Ω, ω	Ω, ω	*Omega*

Index of Symbols

This index lists symbols of physical quantities, units, some mathematical operators, states of aggregation, processes and particles. Symbols of elements are given in Section 6.2 (p.94). Qualifying subscripts, etc., are generally omitted from this index, so that for example E_p for potential energy and E_ea for electron affinity are both indexed simply under E for energy. The Latin alphabet is indexed ahead of the Greek alphabet, lower case letters ahead of upper case, bold symbols ahead of italic, ahead of upright, and single letter symbols ahead of multiletter ones.

a acceleration 11
a fundamental translation vector 36
a* reciprocal lattice vector 36
a absorption coefficient 32
a activity 49, 58
a hyperfine coupling constant 26
a specific surface area 63
a thermal diffusivity 65
a unit cell length 36
a van der Waals coefficient 49
a_0 Bohr radius 20, 76, 89, 110
a adsorbed 47
a are, unit of area 110
a atto, SI prefix 74
a year, unit of time 111
ads adsorbed 47, 51
am amorphous solid 47
amagat amagat unit 113
at atomization 51
atm atmosphere, unit of pressure 54, 89, 112
aq aqueous solution 47

A magnetic vector potential 15
A absorbance 32
A absorption intensity 32
A activity (radioactive) 22
A, $\mathscr{A}$ affinity of reaction 50
A area 11
A Einstein transition probability 30
A Helmholtz energy 48
A hyperfine coupling constant 26
A nucleon number, mass number 20
A pre-exponential factor 56
A rotational constant 23
A spin-orbit coupling constant 23
A van der Waals–Hamaker constant 63
A_H Hall coefficient 37
Al Alfvén number 66
A_r relative atomic mass 41, 94
A ampere, SI unit 71, 114
Å ångström, unit of length 24, 75, 110
AU astronomical unit, unit of length 110

b Burgers vector 36
b fundamental translation vector 36
b* reciprocal lattice vector 36
b breadth 11
b impact parameter 56
b mobility ratio 37
b molality 42
b unit cell length 36
b van der Waals coefficient 49
b barn, unit of area 75, 110
b bohr, unit of length 110
bar bar, unit of pressure 54, 75, 112

B magnetic flux density, magnetic induction 14
B Debye–Waller factor 36
B Einstein transition probability 31
B napierian absorbance 32
B retarded van der Waals constant 63
B rotational constant 23
B second virial coefficient 49
B susceptance 15
B bel, unit of power level 79
Bi biot, unit of electric current 114
Bq becquerel, SI unit 72, 113
Btu British thermal unit, unit of energy 112

c fundamental translation vector 36
c velocity 11, 39
c* reciprocal lattice vector 36
c amount (of substance) concentration 42
c speed 11, 30, 39, 56
c unit cell length 36
c_0 speed of light in vacuum 30, 89
c_1 first radiation constant 32, 89
c_2 second radiation constant 32, 89
c centi, SI prefix 74
c combustion 51
cal calorie, unit of energy 112
cd candela, SI unit 71
cd condensed phase 47
cr crystalline 47

C capacitance 14
C heat capacity 48
C number concentration 39, 42
C rotational constant 23
C third virial coefficient 49
C_n *n*-fold rotation operator 28
C_o Cowling number 66
C coulomb, SI unit 72, 114
Ci curie, unit of radioactivity 113
Cl clausius, unit of entropy 113
°C degree Celsius, SI unit 72, 113

d	centrifugal distortion constant 23
d	collision diameter 56
d	degeneracy 24, 39
d	diameter, distance, thickness 11
d	lattice plane spacing 36
d	relative density 12
d	day, unit of time 75, 111
d	deci, SI prefix 74
d	deuteron 43, 93
da	deca, SI prefix 74
dil	dilution 51
dpl	displacement 51
dyn	dyne, unit of force 112
D	electric displacement 14
D	centrifugal distortion constant 23
D	Debye–Waller factor 36
D	diffusion coefficient 37, 65
D	dissociation energy 20
D_{AB}	direct (dipolar) coupling constant 25
D	debye, unit of electric dipole moment 24, 115
Da	dalton, unit of mass 20, 41, 75, 111
e	unit vector 85
e	elementary charge 20, 58, 76, 89, 114
e	étendue 31
e	linear strain 12
e	base of natural logarithms 84, 90
e	electron 43, 93
erg	erg, unit of energy 112
e.u.	entropy unit 113
eV	electronvolt, unit of energy 75, 112
E	electric field strength 14
E	electric potential difference 58
E	electromotive force 14, 58, 59
E	energy 12, 18–20, 37, 55
E	étendue 31
E	identity symmetry operator 27, 28
E	irradiance 31
E	modulus of elasticity 12
E	thermoelectric force 37
E^*	space-fixed inversion 27
E_h	Hartree energy 20, 76, 89, 112, 120
Eu	Euler number 65
E	exa, SI prefix 74
E	excess quantity 51
f	activity coefficient 50
f	atomic scattering factor 36
f	finesse 31
f	frequency 11
f	friction factor 13
f	fugacity 50
f	vibrational force constant 25
$f(c_x)$	velocity distribution function 39
f	femto, SI prefix 74
f, fm	fermi, unit of length 110
f	formation 51
fl	fluid 47
ft	foot, unit of length 110
fus	fusion 51
F	Fock operator 18, 19
F	force 12
F	total angular momentum 26
F	Faraday constant 58, 89
F	fluence 31
F	rotational term 23
F	structure factor 36
F	vibrational force constant 25
$F(c)$	speed distribution function 39
Fo	Fourier number 65, 66
Fr	Froude number 65
F	farad, SI unit 72
°F	degree Fahrenheit, unit of temperature 113
Fr	franklin, unit of electric charge 114, 118
g	acceleration due to gravity 11, 89, 111
g	degeneracy 24, 39
g	density of vibrational modes 37
g	*g*-factor 21, 26, 89
g	vibrational anharmonicity constant 23
g	gas 47
g	gram, unit of mass 74, 111
gal	gallon, unit of volume 111
gr	grain, unit of mass 111
grad	grade, unit of plane angle 113
G	reciprocal lattice vector 36
G	electric conductance 15
G	Gibbs energy 48, 57
G	gravitational constant 12, 89
G	integrated absorption cross section 33
G	shear modulus 12
G	thermal conductance 65
G	vibrational term 23
G	weight 12
Gr	Grashof number 65, 66
G	gauss, unit of magnetic flux density 115
G	giga, SI prefix 74
Gal	gal, unit of acceleration 111
Gy	gray, SI unit 72, 113
h	coefficient of heat transfer 65
h	film thickness 63
h	height 11
h	Miller index 38
h, $\hbar$	Planck constant ($\hbar = h/2\pi$) 20, 30, 76, 89
h	hecto, SI prefix 74
h	helion 43, 93
h	hour, unit of time 75, 111
ha	hectare, unit of area 110
hp	horse power, unit of power 112
H	magnetic field strength 14
H	enthalpy 48, 56
H	fluence 31
H	Hamilton function 12, 16
Ha	Hartmann number 66
H	henry, SI unit 72
Hz	hertz, SI unit 11, 72
i	unit vector 85
i	electric current 14
i	inversion operator 28
i	square root of -1 85
id	ideal 51
imm	immersion 51
in	inch, unit of length 110
I	nuclear spin angular momentum 26
I	differential cross section 56

Σ	spin component quantum number	26
Σ_f	film tension	63
Σ	summation sign	84

τ	acoustic transmission factor	13
τ	characteristic time, relaxation time	11, 22, 37, 55
τ	chemical shift	25
τ	shear stress	12
τ	thickness of surface layer	63
τ	Thomson coefficient	37
τ	transmittance	32

ϕ	electric potential	14
ϕ	fluidity	13
ϕ	fugacity coefficient	50
ϕ	inner electric potential	59
ϕ	molecular orbital	17, 18, 19
ϕ	osmotic coefficient	51
ϕ	plane angle	11
ϕ	quantum yield	57
ϕ	spherical coordinate	11
ϕ_{rst}	vibrational force constant	25
ϕ	volume fraction	41
ϕ	wavefunction	16

Φ	heat flow rate	65
Φ	magnetic flux	14
Φ	potential energy	12
Φ	quantum yield	57
Φ	radiant power	31
Φ	work function	37

χ	quadrupole interaction energy tensor	22
χ	atomic orbital	17, 19
χ	electronegativity	20
χ	magnetic susceptibility	15
χ	surface electric potential	59

χ_e	electric susceptibility	14
χ_m	molar magnetic susceptibility	15

ψ	outer electric potential	59
ψ	wavefunction	16

Ψ	electric flux	14
Ψ	wavefunction	16, 18

ω	harmonic vibration wavenumber	23
ω	angular frequency, angular velocity	11, 21, 30, 36
ω	statistical weight	39
ω	solid angle	11

Ω	angular momentum component quantum number	26
Ω	partition function	39
Ω	solid angle	11
Ω	volume in phase space	39
Ω	ohm	72, 114

Special symbols

%	percent	77
‰	permille	78
°	degree, unit of arc	75, 113
°	standard	51
$\ominus$	standard	51
′	minute, unit of arc	75, 113
″	second, unit of arc	75, 113
*	complex conjugate	16, 85
*	excitation	44
*	pure substance	51
‡	activation, transition state	51, 56
∞	infinite dilution	51
[B]	concentration of B	42
$[\alpha]$	specific optical rotatory power	33

Subject Index

When more than one page reference is given, bold print is used to indicate the most useful general reference. Greek letters are spelled out and accents are ignored in alphabetical ordering.

NOTES

NOTES

NOTES

NOTES

NOTES

PRESSURE CONVERSION FACTORS

	Pa	kPa	bar	atm	Torr	psi
1 Pa =	1	10^{-3}	10^{-5}	$9.869\,23 \times 10^{-6}$	$7.500\,62 \times 10^{-3}$	$1.450\,38 \times 10^{-4}$
1 kPa =	10^3	1	10^{-2}	$9.869\,23 \times 10^{-3}$	7.500 62	0.145 038
1 bar =	10^5	10^2	1	0.986 923	750.062	145.038
1 atm =	101 325	101.325	1.013 25	1	760	14.6959
1 Torr =	133.322	0.133 322	$1.33\,322 \times 10^{-3}$	$1.315\,79 \times 10^{-3}$	1	$1.933\,67 \times 10^{-2}$
1 psi =	6894.76	6.894 76	$6.894\,76 \times 10^{-2}$	$6.804\,60 \times 10^{-2}$	51.715 07	1

Examples of the use of this table:

1 bar = 0.986 923 atm

1 Torr = 133.322 Pa

Note: 1 mmHg = 1 Torr, to better than 2×10^{-7} Torr (see p.112).